Mechanical Symmetry

by

Joaquín Obregón Cobo

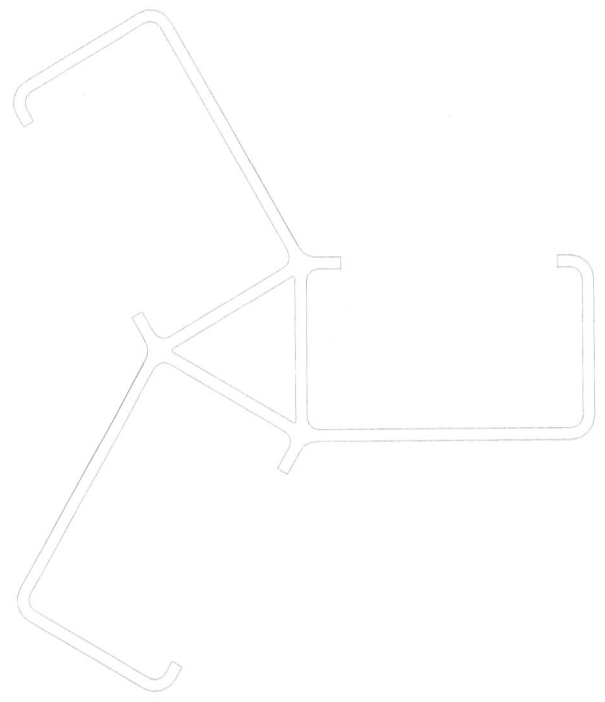

authorHOUSE®

AuthorHouse™
1663 Liberty Drive
Bloomington, IN 47403
www.authorhouse.com
Phone: 1-800-839-8640

© 2012 by Joaquín Obregón Cobo. All rights reserved.

No part of this book may be reproduced, stored in a retrieval system, or transmitted by any means without the written permission of the author.

Published by AuthorHouse   12/10/2012

ISBN: 978-1-4772-3372-6 (sc)
ISBN: 978-1-4772-4573-6 (hc)
ISBN: 978-1-4772-4574-3 (e)

Library of Congress Control Number: 2012923300

This book is printed on acid-free paper.

Because of the dynamic nature of the Internet, any web addresses or links contained in this book may have changed since publication and may no longer be valid. The views expressed in this work are solely those of the author and do not necessarily reflect the views of the publisher, and the publisher hereby disclaims any responsibility for them.

In Internet:
http://www.mecsym.org   and   http://www.mechanicalsymmetry.com

Versión Española disponible. ISBN: 978-1-4772-3117-3

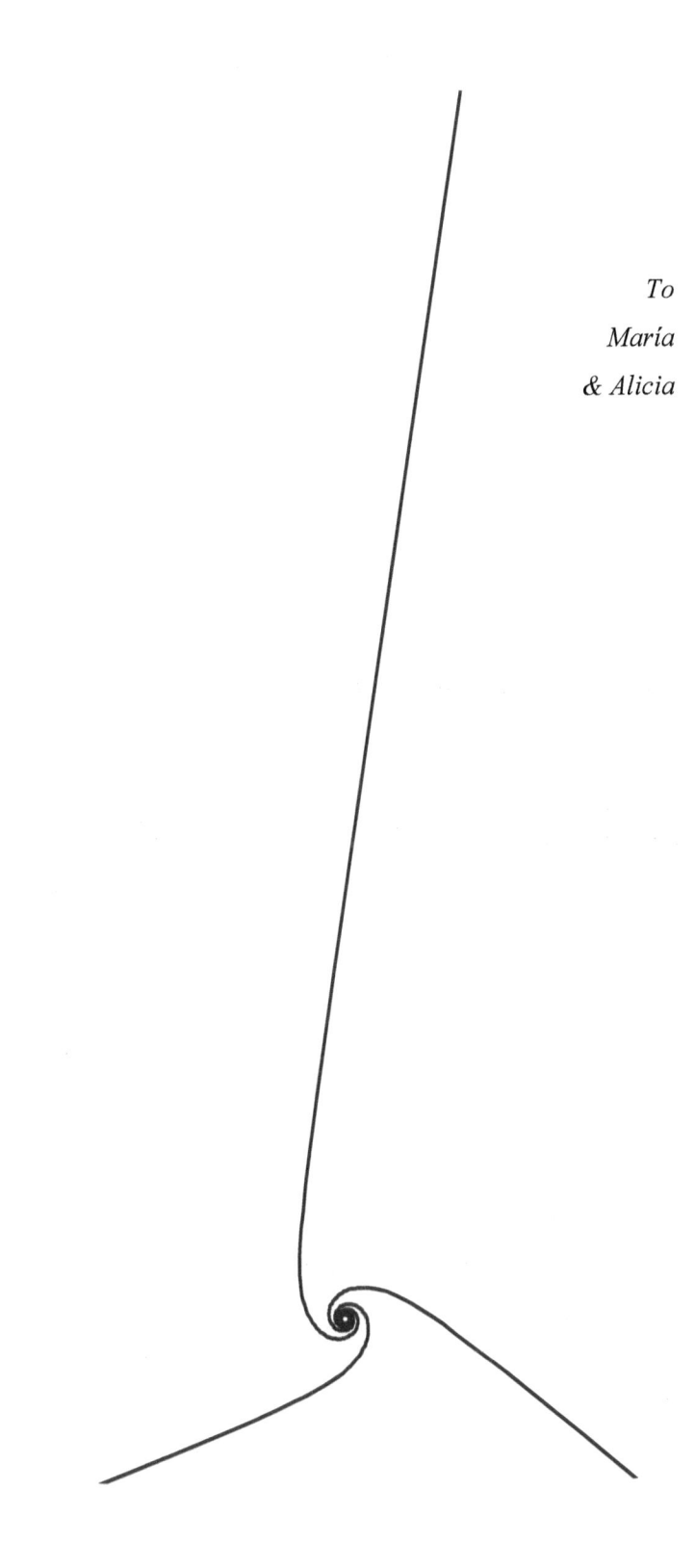

*To
María
& Alicia*

Index

1. Introduction	3
1.1 Moments	3
1.2 Moment of Inertia (MoI)	4
1.3 Steiner's Theorem	4
1.4 Symmetry	5
1.E Exercise about MoI	8
2 Questions and Known Answers	17
2.1 What Happens if We Rotate a Section?	17
2.2 Can I Have a Formula to Calculate Rotated MoI?	17
2.3 Can We Simplify MoI Calculation when Rotating Sections?	18
2.4 Steiner's Theorem in Detail	18
2.5 MoI Formulas for Rotation	21
2.6 Superposition	23
2.E Exercise - MoI Calculation with Translation and Rotation	25
3 New Answers — Mechanical Symmetry	35
3.1 Previous Test	35
3.2 First Approach	37
3.3 Generalization	45
3.E Exercises about Mechanical Symmetry	47
3.4 Definition and Theorem	51
3.5 Corollary	51
3.6 Necessity and Sufficiency	51
3.7 Theorem	58
3.8 Why Should We Do It?	61
4 No Answers	65
4.1 Incoherence	65
4.2 Incoherent Sums	66
5. Using It	71
5.1 Regular Polygons	71
5.2 Circles and Computers	75
6. Formula Compilation	91
6.1 Sections with Mechanical Symmetry	91
6.2 Regular Polygons	95
6.3 Regular Polygonal Tubes	113

6.4 Regular Polygonal Stars	127
Appendix 1: Computer Programs for Chapter 2	**137**
Ap. 1.1 Showing MoI Changing with Rotation and Translation	139
Appendix 2: Computer Programs for Chapter 3	**151**
Ap. 2.2 Calculation of sine² Sums	153
Ap. 2.2 Drawing the sine² Sums	157
Ap. 2.3 Particles MoI Comparison	165
Ap. 2.4 Particles MoI Comparison – Table	169
Ap. 2.5 MoI Graphical Interactive Comparison	175
Appendix 3: Computer Programs for Chapter 5	**191**
Ap. 3.1 Regular Polygons MoI Interactive Calculation	193
Ap. 3.2 Regular Polygonal Tubes MoI Interactive Calculation	207
Ap. 3.3 Regular Polygon Based Stars MoI Interactive Calculation	211
Appendix 4: MoI Calculation for any Polygon	**215**
Ap. 4.1 Trapezoid MoI Sums	217
Ap. 4.2 Green Theorem	218
Index of Figures	225
Index of Equations	227
Index of Tables	229
Glossary	231
Bibliography	233
Acknowledgements	235

Introduction

Mechanical Symmetry

Mechanical Symmetry

1. Introduction

The moment of inertia (MoI), also known as second moment of area, is not the most innovative concept today; it is a long time since the basis of it was firmly settled. But MoI is used every day, and we see it is interesting due to the extensive use it has.

One may ask: is there anything interesting about MoI? The answer is YES, and one will find it in this book. We will see that MoI can be constant under certain conditions, potentially simplifying many calculations. We will also see some new formulas that will allow us to express algebraically some problems that are limited now to tables and abacuses.

Now for a short introduction to the concepts used in the book.

1.1 Moments

A moment is a magnitude related to an axis and to a given property. We can define moments related to a point or to a plane, too, but we will not do so in this book.

The moment related to an axis is defined as the product of the value of a property in a point times the distance from the point to the axis. This will be the first moment or simply moment. Second moment is the product of the value of the property times the square of the distance to the axis. This is also known as second-order moment. We can define the n-order moment as the product of the property's value times the n-power of the distance to the axis.

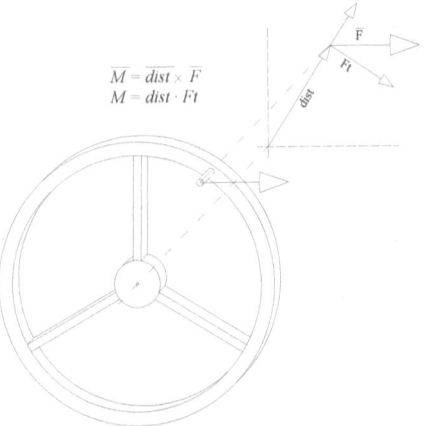

Fig. 1 Torque

Torque is a good example to understand what a moment is. The physical property is the force (only the tangential component Ft, to be more precise), and the distance is the distance from the force to the axis.

$$\vec{M} = \overrightarrow{dist} \times \vec{F} \quad \Rightarrow \quad M = dist \cdot Ft$$

Eq. 1 Torque

1.2 Moment of Inertia (MoI)

The moment of inertia of a given section (usually a structural section) is the second moment of area. That is, the product of the area in the points of the section times the square of the distance from the points to the axis. We will abbreviate it MoI, and it will appear as I in formulas.

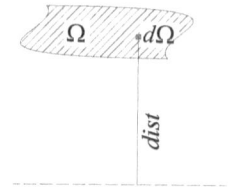

Fig. 2 Solid Moment of Inertia

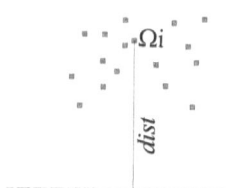

Fig. 3 Particles Moment of Inertia

Ω is the total area and $d\Omega$ is an area infinitesimal.

Ω_i is the area for particle i.

$$I = \int_{\Omega} dist^2 \cdot d\Omega$$

Eq. 2 Solid Moment of Inertia

$$I = \sum_{i=1}^{n} dist^2 \cdot \Omega_i$$

Eq. 3 Particles Moment of Inertia

1.3 Steiner's Theorem

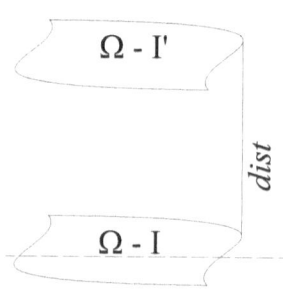

Fig. 4 Steiner's Theorem

Given the MoI related to an axis through the center of mass of a section, if we want to know the MoI related to a different axis parallel to the first one; Mr. Steiner will help us.

According to Steiner's theorem, the value of the new MoI equals the previous one, plus the product of the square of the distance from the center of mass to the new axis times the total area.

$$I' = I + \Omega \cdot dist^2$$

Eq. 4 Steiner's Theorem

Mechanical Symmetry

1.4 Symmetry

What does symmetry mean? From the ancient Greek, the direct translation is "paired measure." For us, symmetry will have two different meanings:
• When we are dealing with physical properties (as MoI is), symmetry means that we have a constant value for a property after making a transformation or movement.
• When we are dealing with geometry, the meaning is that it keeps its shape exactly the same after making any transformation or movement.

We are going to develop our exposition in 2D plane, and then the relevant geometric symmetry types are:

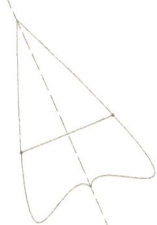
Fig. 5 Axial Symmetry

Fig. 6 Point Symmetry

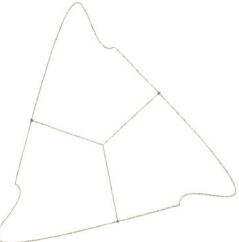
Fig. 7 Rotational Symmetry

Axial symmetry is the most widely known symmetry, and it is the one we could say is the most recognizable. It is also known as specular symmetry. This symmetry's defining characteristic is that every point has a corresponding symmetric point on the other end of a segment perpendicular to the line defining the symmetry, also known as the symmetry axis.

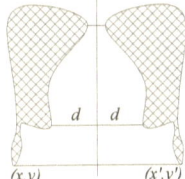  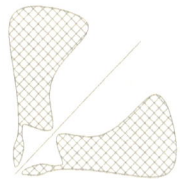

Fig. 8 Axial Symmetry - Figures

From a mathematical point of view, we can express this symmetry as:

$$x' = -x$$
$$y' = y$$

Obviously this is for the case of the symmetry axis is the vertical axis y=0. If the symmetry axis is a different line, we can always apply a

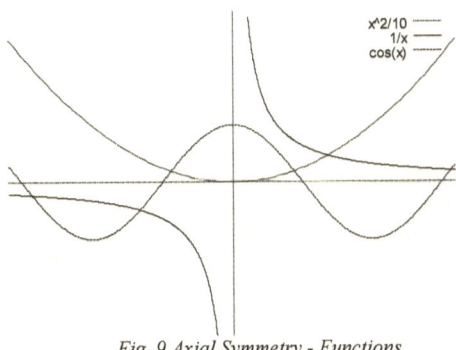

Fig. 9 Axial Symmetry - Functions

translation and a rotation in order to make the symmetry axis be the vertical axis y=0.

In *Fig. 9 Axial Symmetry - Functions* we can see some samples of functions with axial symmetry, also known as even functions:

$$y = \frac{x^2}{10} \quad y = \frac{1}{x} \quad y = \cos x$$

This type of symmetry is frequently found in nature and in some arts and artistries.

Point symmetry is neither widely known nor easy to recognize. In this type of symmetry, the symmetric points are on the ends of segments, with their midpoint in the point defining the symmetry, known as symmetry center.

From a mathematical point of view, we can express this symmetry as:

$$x' = -x$$
$$y' = -y$$

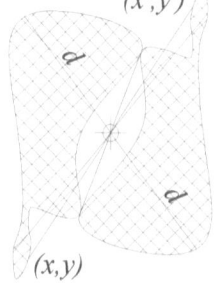

Fig. 10 Point Symmetry - Figure

Obviously this is when the symmetry center is the origin, or x=0, y=0. If the symmetry center is a different point, we can always apply a translation in order to make the symmetry center be the origin (0, 0).

Some functions have point symmetry, such as the functions in *Fig. 11 Point Symmetry - Functions*.

$$y = \frac{x^3}{10} \quad y = \frac{1}{x} \quad y = \sin x$$

We also see that some functions have both symmetries, like the hyperbolic one, which is symmetric with respect to:

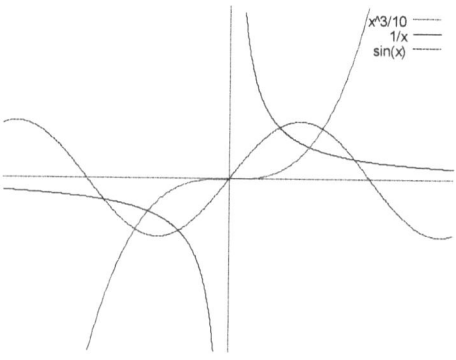

Fig. 11 Point Symmetry - Functions

$$y = x \quad y = -x \quad (0,0)$$

Mechanical Symmetry

Rotational symmetry is the least known, and it is sometimes confused with point symmetry. In this type of symmetry, there are *k* symmetrical points located at *k* evenly spaced intervals along a circumference with its center in the point defining the symmetry or center of symmetry.

That's why we need the **order *k*** to completely define rotational symmetry.

In *Fig. 12*, figures with rotational symmetries of orders 2, 3, 4 and 5 are shown, and figures with 2, 3 and 4 order are in *Table 5 Fourier – Rotational Symmetry Samples*.

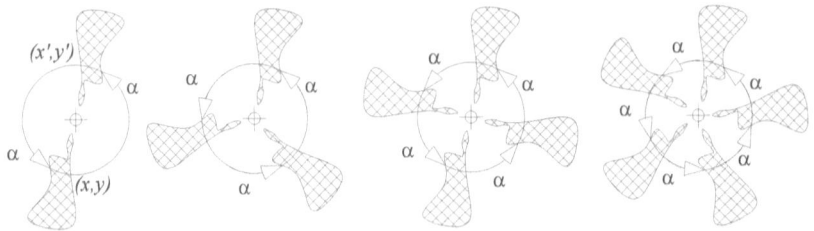

Fig. 12 Rotational Symmetry - Figures

From a mathematical point of view, we can express this symmetry as:

$$x = \rho\cos\theta \quad x_n' = \rho\cos(\theta + n\tfrac{2\pi}{k})$$
$$y = \rho\sin\theta \quad y_n' = \rho\sin(\theta + n\tfrac{2\pi}{k})$$

$$\rho = \sqrt{x^2 + y^2} \quad x_n' = \sqrt{x^2 + y^2}\cos(arctg\tfrac{y}{x} + n\tfrac{2\pi}{k})$$
$$\theta = arctg\tfrac{y}{x} \quad y_n' = \sqrt{x^2 + y^2}\sin(arctg\tfrac{y}{x} + n\tfrac{2\pi}{k})$$

Using polar coordinates it is:

$$\theta' = \theta + n\tfrac{2\pi}{k}$$

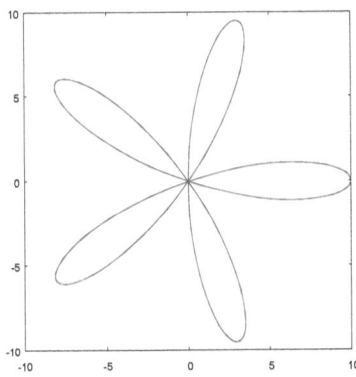

Fig. 13 Function with Rotational Symmetry

As before, the symmetry center is located at the origin of coordinates, etc.

In *Fig. 13* we see the function with rotational symmetry:

$$\rho = \sin 5 \cdot \theta$$

It is interesting that rotational symmetry is present in many natural processes related to growing.

Nature also shows to us that rotational symmetry is not limited to a triangulation of the circle (i.e., to a division of it in equal triangles).

Mechanical Symmetry

1.E Exercise about MoI

As a practice on the basics of the moment of inertia, we are going to calculate the MoI of some commonly used figures. We are going to do it with an example run in "Maxima" software, accompanied by some illustrative figures.

---

1. Rectangle

1. Ix

A Little bit of cleaning (cleans the working environment).
(%i10) kill (all);
(%o0) *done*

First we calculate the MoI for a vertical infinitesimal with height h.
(%i1) da:integrate(y^2, y,-h/2,h/2);

(%o1) $\dfrac{h^3}{12}$

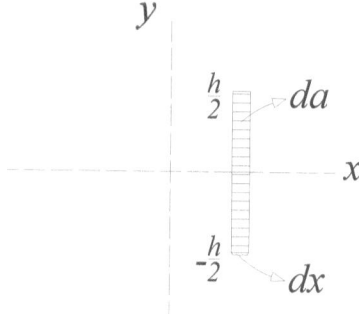

Then we integrate this MoI infinitesimal:
(%i2) I:integrate(da, x, -b/2, b/2);

(%o2) $\dfrac{bh^3}{12}$

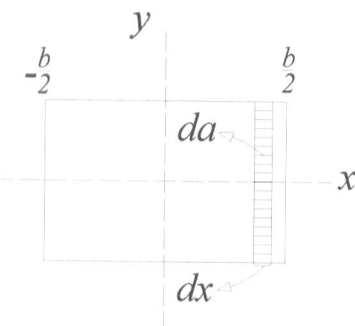

2. Iy

First we calculate the MoI for a horizontal infinitesimal with width b:
(%i3) da:integrate(y^2, y,-b/2,b/2);

(%o3) $\dfrac{b^3}{12}$

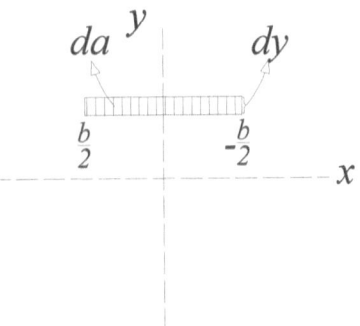

Then we integrate this MoI infinitesimal:
(%i4) I:integrate(da, x, -h/2, h/2);

(%o4) $\dfrac{b^3 h}{12}$

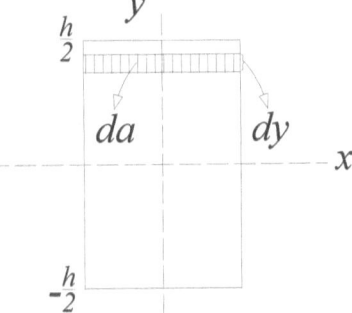

---

2. Triangle

Isosceles triangle with vertex at origin.

1. Ix

(%i5) kill(all);
(%o0) *done*

Triangle side equation:
(%i1) y(x):=x*(h/2)/b;

(%o1) $y(x) := \dfrac{x \dfrac{h}{2}}{b}$

Mechanical Symmetry

```
(%i2) y(b);
```
$$(\%o2)\frac{h}{2}$$

```
(%i3) y(0);
```
(%o3)0

Calculate the MoI for a vertical infinitesimal with height y(x):
```
(%i4) da:integrate(h^2, h,-y(x),y(x));
```
$$(\%o4)\frac{h^3 x^3}{12 b^3}$$

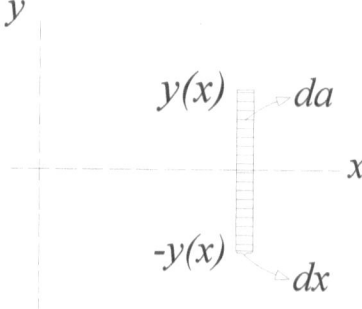

Integrate this MoI infinitesimal from 0 to b:
```
(%i5) I:integrate(da, x,0,b);
```
$$(\%o5)\frac{b h^3}{48}$$

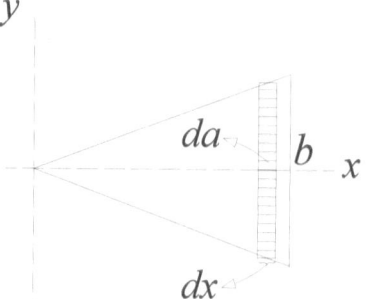

2. Iy

```
(%i6) kill(all);
```
(%o0)*done*

Triangle side equation:
```
(%i1) x(y):=y/(h/2)*b;
```
$$(\%o1) x(y) := \frac{y}{\frac{h}{2}} b$$

```
(%i2) x(0);
(%o2) 0

(%i3) x(h/2);
(%o3) b
```

MoI for an horizontal infinitesimal from x(y) to b:
```
(%i4) da:integrate(j^2, j,x(y),b);
```

(%o4) $\dfrac{b^3}{3} - \dfrac{8b^3 y^3}{3h^3}$

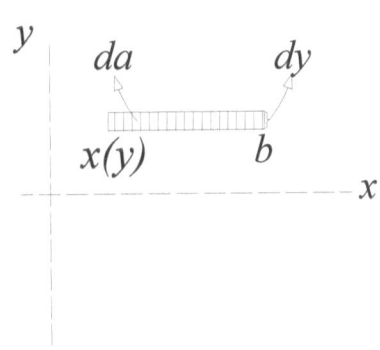

Integrate this MoI infinitesimal from 0 to h/2:
```
(%i5) I:2*integrate(da, y,0,h/2);
```

(%o5) $\dfrac{b^3 h}{4}$

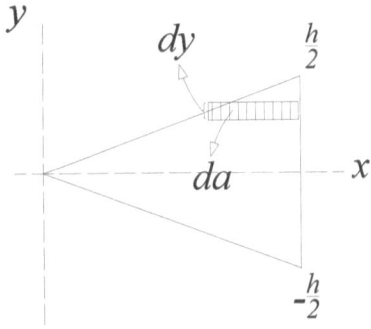

Steiner to get MoI from Center of Mass:
```
(%i6) Icdg:I-(b*h/2)*(2*b/3)^2;
```

(%o6) $\dfrac{b^3 h}{36}$

3. Circle

1. Polar

Preparation for clean calculation:
(%i7) kill(all);
(%o0)*done*

(%i1) assume(r>0);
(%o1)$[r>0]$

Distance to axis for the center of mass of the area infinitesimal:
(%i2) y:r*sin(beta)/2;
(%o2)$\dfrac{sin(beta)r}{2}$

Area infinitesimal value: a=r*r*dbeta. Note dbeta is not written' it is implicit in the integration.
(%i3) a:r*r;
(%o3)r^2

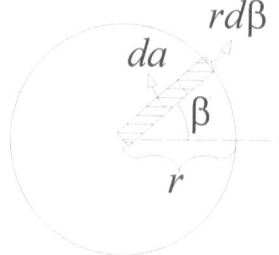

Integrate Fint*da:
(%i4) Fint:y^2*a;
(%o4)$\dfrac{sin(beta)^2 r^4}{4}$

(%i5) I:integrate(Fint, beta,0,2*%pi);
(%o5)$\dfrac{\pi r^4}{4}$

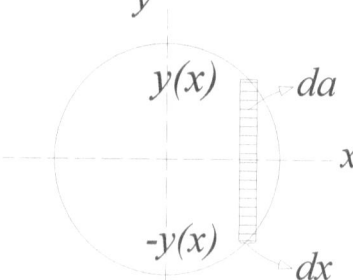

2. Cartesian

```
(%i6) kill(y);
```
(%o6) *done*

First we calculate the MoI for a vertical infinitesimal.
```
(%i7) da:integrate(y^2, y,-y,y);
```
$$(\%o7)\, \frac{2y^3}{3}$$

Circle equation:
```
(%i8) y:sqrt(r^2-x^2);
```
$$(\%o8)\, \sqrt{r^2 - x^2}$$

Integrate Fint*dx:
```
(%i9) Fint:ev(da, nouns);
```
$$(\%o9)\, \frac{\sqrt{2(r^2-x^2)^3}}{3}$$

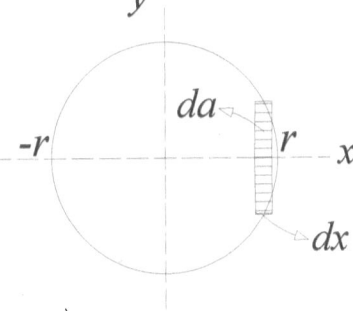

```
(%i10) I:integrate(Fint, x,-r,r);
```
$$(\%o10)\, \frac{\pi r^4}{4}$$

Mechanical Symmetry

Known Answers 2

Mechanical Symmetry

2 Questions and Known Answers

We have questions, and we are looking for answers for our real, physical world, having in mind structures and elements with physical properties and not allowing geometric transformations like inversions and scaling; our real-world entities will be translated and/or rotated. Consequently, we will deal only with translations and rotations in this book. Also, Steiner's theorem solves any possible question regarding translations, so we will concentrate on rotations.

2.1 What Happens if We Rotate a Section?

The MoI will change. Only in some special cases the MoI will be the same before and after rotation:

a) For sections with constant MoI. This is a small set (it will be bigger at the end of the book), including round sections and some others.
b) Rotating 180 degrees.
c) For sections with k-order rotational symmetry rotating $\alpha=360/n$ degrees, or any multiple of α.

2.2 Can I Have a Formula to Calculate Rotated MoI?

As a general rule, we can say that after rotating by an angle α, the value of the MoI will be:

$$I_x^\alpha = I_x \cos^2 \alpha + I_y \sin^2 \alpha - I_{xy} \sin 2\alpha$$

$$I_y^\alpha = I_x \sin^2 \alpha + I_y \cos^2 \alpha + I_{xy} \sin 2\alpha$$

$$I_{xy}^\alpha = \frac{I_x - I_y}{2} \sin 2\alpha + I_{xy} \cos 2\alpha$$

Eq. 5 Moment of Inertia – Rotation

We see something new here. Yes, MoI never walks alone. We see that every MoI I_x has its orthogonal companion I_y and the I_{xy} to completely define the MoI of the section in any direction.

As a conclusion: we need "a lot of" information and "quite complex" formulas to be able to handle MoI value when rotating a section.

2.3 Can We Simplify MoI Calculation when Rotating Sections?

No but ...

We cannot simplify the calculations, but we can "simplify" our section to get simpler formulas.

From the previous point, we see that we can find the angle that makes $I_{xy} = 0$. This angle defines the principal directions and moments of inertia I_u and I_v.

$$I_{xy}{}^\alpha = \frac{I_x - I_y}{2}\sin 2\alpha + I_{xy}\cos 2\alpha = 0 \Rightarrow \alpha = -\tfrac{1}{2}arctg\frac{2I_{xy}}{I_x - I_y}$$

Eq. 6 Moment of Inertia - Principal Directions

But this only gives to us the principal directions. If we want to simplify, we need to have equal principal MoI I_u and I_v arriving to:

$$I_x{}^\alpha = \frac{I_u + I_v}{2} + \frac{I_u - I_v}{2}\cos 2\alpha = I$$

$$I_y{}^\alpha = \frac{I_u + I_v}{2} - \frac{I_u - I_v}{2}\cos 2\alpha = I$$

$$I_{xy}{}^\alpha = \frac{I_u - I_v}{2}\sin 2\alpha = 0$$

$$\forall \alpha \in \Re$$

Eq. 7 Constant Moment of Inertia

In other words, what we need is a section with constant MoI. For such a section, we say that the ellipse of inertia defined by the values of I_x and I_y is a circle:

$$I_x{}^\alpha = I_x = I_y = I = \text{constant} \quad (I_{xy} = 0)$$

It is a section with no dependence on the angle for the MoI.

2.4 Steiner's Theorem in Detail

The algebraic expression for the theorem is well-known:

$$I' = I + \Omega \cdot \delta^2$$

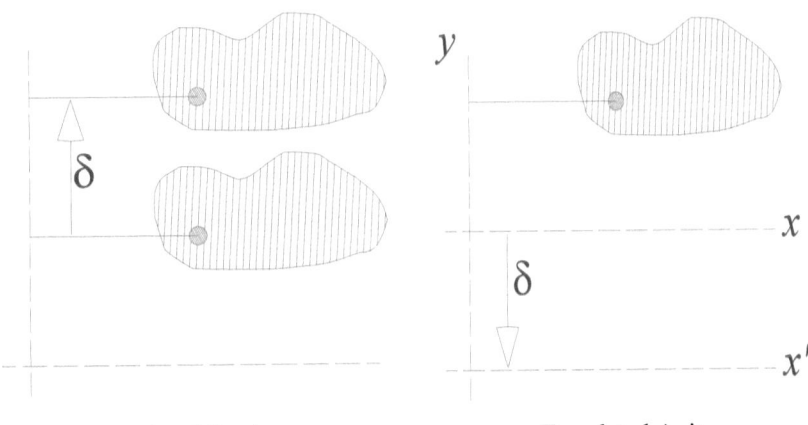

Translated Section   Translated Axis

Fig. 14 Steiner - Section and Axis Translation

This can be verified both for the translation of the section and for the opposite movement applied to the axis. Point coordinates after the translation are:

$$x' = x$$
$$y' = y + \delta$$

With the new coordinates, we integrate the formula defining the MoI:

$$I = \int_\Omega dist^2 \cdot d\Omega$$

$$I' = \int_\Omega (y')^2 \cdot d\Omega = \int_\Omega (y+\delta)^2 d\Omega = \int_\Omega (y^2 + \delta^2 + 2\delta y) d\Omega =$$
$$= \int_\Omega y^2 d\Omega + \delta^2 \int_\Omega d\Omega + 2\delta \int_\Omega y d\Omega = I + \Omega \delta^2 + 2\delta \int_\Omega y d\Omega$$

To arrive to Steiner's formula, we need

$$2\delta \int_\Omega y d\Omega = 0$$

Steiner's theorem is for an axis through the center of mass, and for this point.

$$\int_\Omega y d\Omega = 0$$

Proving that $I' = I + \Omega \cdot \delta^2$ for an axis located at the center of mass.

Mechanical Symmetry

What happens to I_{xy} after a translation?

$$I'_{xy} = \int_\Omega x \cdot y' \cdot d\Omega = \int_\Omega x(y+\delta) d\Omega = \int_\Omega (xy+x\delta) d\Omega =$$

$$= \int_\Omega xy\, d\Omega + \delta \int_\Omega x\, d\Omega = I_{xy} + \delta \int_\Omega x\, d\Omega$$

As far as Steiner's theorem refers to axis through the center of mass:

$$\int_\Omega x\, d\Omega = 0$$

That proves that I_{xy} does not change ($I'_{xy} = I_{xy}$) after a translation from an axis passing through the center of mass.

A common mistake, which may be caused by a misinterpretation of previous figures, is to calculate MoI from a value of the MoI not corresponding to the center of mass. Observe the following figures.

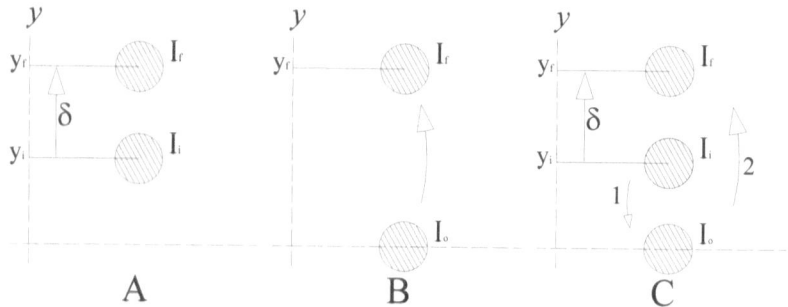

Fig. 15 Steiner - Correct calculation - Axis not at center of mass

If, as shown in figure A, we calculate I_f with Steiner's formula from I_i, then we will get a wrong result:

$$NO!\ \ \cancel{I_f = I_i + \Omega \delta^2}\ \ [I]$$

We know that the correct value is calculated, as shown in figure B, calculating I_f with Steiner's formula from I_o, that is, with a MoI value passing through the center of mass.

$$I_f = I_o + \Omega y_f^2 = I_o + \Omega(y_i + \delta)^2$$

And we know, because it is the exact expression of Steiner's theorem

$$I_i = I_o + \Omega y_i^2$$

By substituting in [1]

$$I_f = \left(I_o + \Omega y_i^2\right) + \Omega \delta^2 = I_o + \Omega(y_i^2 + \delta^2) \neq I_o + \Omega(y_i + \delta)^2$$

we see that [1] is incorrect.

We have to proceed as shown in figure C:

First calculate I_o and then I_f

$$I_o = I_i - \Omega y_i^2$$

$$I_f = I_o + \Omega y_f^2 = I_o + \Omega(y_i + \delta)^2 = \left(I_i - \Omega y_i^2\right) + \Omega(y_i + \delta)^2$$

$$I_f = I_i + \Omega(2\delta y_i + \delta^2)$$

Arriving to a formula for I_f as a function of I_i and δ, but MoI depends also on y_i in this case. It is not possible to get it directly from I_i and δ.

2.5 MoI Formulas for Rotation

As a general rule, we can say that after rotating by an angle α, the value of the MoI will be:

$$I_x^\alpha = I_x \cos^2 \alpha + I_y \sin^2 \alpha - I_{xy} \sin 2\alpha$$

$$I_y^\alpha = I_x \cos^2 \alpha + I_y \sin^2 \alpha + I_{xy} \sin 2\alpha$$

$$I_{xy}^\alpha = \frac{I_x - I_y}{2} \sin 2\alpha + I_{xy} \cos 2\alpha$$

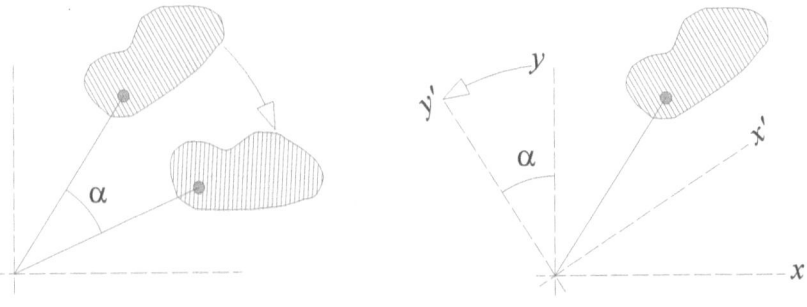

Rotated Section   Rotated Axis

Fig. 16 MoI Rotation – Section and Axis Rotation

Mechanical Symmetry

Formulas that we can obtain by rotating the reference axis to which MoI is calculated an angle α, is exactly the same as rotating the section the same angle in the opposite direction. Point coordinates before and after rotation are as follows.

$$x = \rho\cos\theta \quad x' = \rho\cos(\theta - \alpha)$$
$$y = \rho\sin\theta \quad y' = \rho\sin(\theta - \alpha)$$

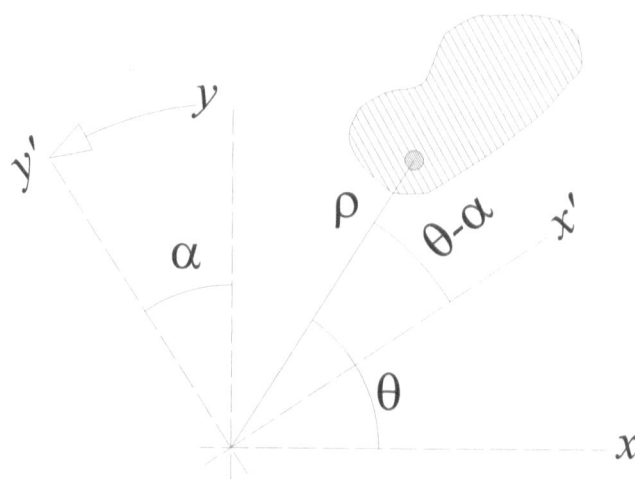

Fig. 17 Section Rotation Formulas

$$x' = \rho\big(\cos(\alpha)\cos(\theta)+\sin(\alpha)\sin(\theta)\big) = x\cos\alpha + y\sin\alpha$$
$$y' = \rho\big(-\sin(\alpha)\cos(\theta)+\cos(\alpha)\sin(\theta)\big) = -x\sin\alpha + y\cos\alpha$$

With the new coordinates, we integrate the formula defining the MoI.

$$I = \int_\Omega dist^2 \cdot d\Omega$$

$$I_x^\alpha = \int_\Omega (y')^2 \cdot d\Omega = \int_\Omega (-x\sin\alpha + y\cos\alpha)^2 d\Omega =$$

$$= \int_\Omega (x^2 \sin^2\alpha + y^2 \cos^2\alpha - 2xy\cos\alpha\sin\alpha) d\Omega =$$

$$= \cos^2\alpha \int_\Omega y^2\, d\Omega + \sin^2\alpha \int_\Omega x^2\, d\Omega - 2\cos\alpha\sin\alpha \int_\Omega xy\, d\Omega =$$

$$= I_x \cos^2\alpha + I_y \sin^2\alpha - I_{xy} 2\cos\alpha\sin\alpha =$$

$$= I_x \cos^2\alpha + I_y \sin^2\alpha - I_{xy} \sin 2\alpha = I_x^\alpha$$

$$I_y^\alpha = \int_\Omega (x')^2 \cdot d\Omega = \int_\Omega (x\cos\alpha + y\sin\alpha)^2 d\Omega =$$

$$= \int_\Omega (x^2 \cos^2\alpha + y^2 \sin^2\alpha + 2xy\cos\alpha\sin\alpha) d\Omega =$$

$$= \cos^2\alpha \int_\Omega x^2 d\Omega + \sin^2\alpha \int_\Omega y^2 d\Omega + 2\cos\alpha\sin\alpha \int_\Omega xy\, d\Omega =$$

$$= I_y \cos^2\alpha + I_x \sin^2\alpha + I_{xy} 2\cos\alpha\sin\alpha =$$

$$= I_x \sin^2\alpha + I_y \cos^2\alpha + I_{xy} \sin 2\alpha = I_y^\alpha$$

$$I_{xy}^\alpha = \int_\Omega x'y' d\Omega = \int_\Omega (x\cos\alpha + y\sin\alpha)(-x\sin\alpha + y\cos\alpha) d\Omega =$$

$$= \int_\Omega (-x^2 \sin\alpha\cos\alpha + xy\cos^2\alpha - xy\sin^2\alpha + y^2 \sin\alpha\cos\alpha) d\Omega =$$

$$= \int_\Omega \left(y^2 \frac{\sin 2\alpha}{2} - x^2 \frac{\sin 2\alpha}{2} + xy\cos 2\alpha \right) d\Omega =$$

$$= \frac{\sin 2\alpha}{2} \left(\int_\Omega y^2 d\Omega - \int_\Omega x^2 d\Omega \right) + \cos 2\alpha \int_\Omega xy\, d\Omega =$$

$$= (I_x - I_y) \frac{\sin 2\alpha}{2} + I_{xy} \cos 2\alpha =$$

$$= \frac{(I_x - I_y)}{2} \sin 2\alpha + I_{xy} \cos 2\alpha = I_{xy}^\alpha$$

2.6 Superposition

Some text cites superposition as a theorem, but we simply consider it a direct consequence of the concept's definition: the total MoI of a given section can be obtained as the sum of the parts by which it is composed. To show it, we rewrite the concept definition.

$$I_{Total} = I_1 + I_2 + \ldots + I_n = \sum_{i=1}^{n} I_i$$

With the following example, we will show superposition and also the influence of distance to axis on the value of MoI. Let's consider a section composed by two rectangles with width b and height h, and with their centers of mass located at a distance h from the axis.

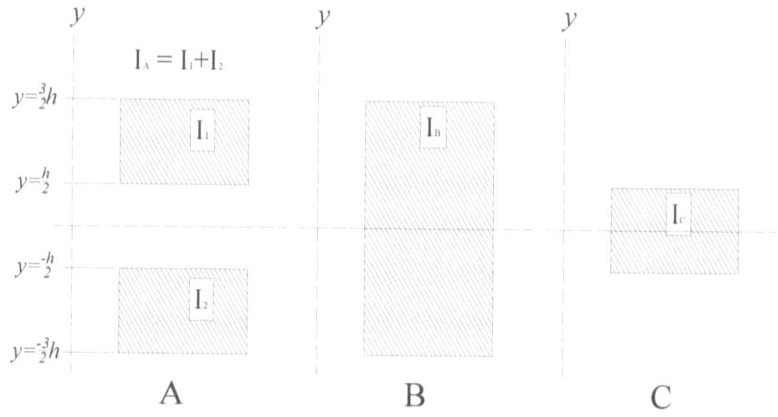

Fig. 18 Superposition and Distance to Axis Influence on MoI

$$I_{Total} = I_A = I_1 + I_2$$

$$I_1 = I_2 = \frac{bh^3}{12} + \Omega \cdot dist^2 = \frac{bh^3}{12} + bh \cdot h^2 = \frac{13}{12}bh^3$$

$$I_A = 2\frac{13}{12}bh^3 = \frac{26}{12}bh^3$$

$$I_B = \frac{b \cdot (3h)^3}{12} = \frac{27}{12}bh^3$$

$$I_C = \frac{1}{12}bh^3$$

$$I_B - I_C = \frac{27}{12}bh^3 - \frac{1}{12}bh^3 = \frac{26}{12}bh^3 = I_A$$

This example proves superposition, and we can also see, paying attention to ratios between figures A, B and C, that 2/3 of the weight (A area against B area) gives 26 times more MoI or in others words, that the third of the area (weight) around the axis is only giving a 1/27 part of the MoI (in section B). This is the argument that leads us to associate the distance of the area of a section to the (neutral) axis with a higher MoI. For example, standardized I sections are the practical materialization of this approach, in the same way the location of the steel reinforcement in the concrete near the external perimeter of the section maximizes the efficiency of the section.

2.E Exercise - MoI Calculation with Translation and Rotation

Now we are going to show all this with an exercise; we will also compare the results obtained from the Steiner application from this book (Appendix 1) against the results from a workbook in mathematical software package Maxima.

We define a circle and a rectangle with the same area.

```
(%i1)  r:2;
```
(%o1)2

1. Area
1. Circle
```
(%i2)  Ac:%pi*r^2;
```
(%o2)4π

2. Rectangle
```
(%i3)  b:r;
```
(%o3)2

```
(%i4)  h:%pi*r;
```
(%o4)2π

```
(%i5)  Ar:b*h;
```
(%o5)4π

2. Initial position

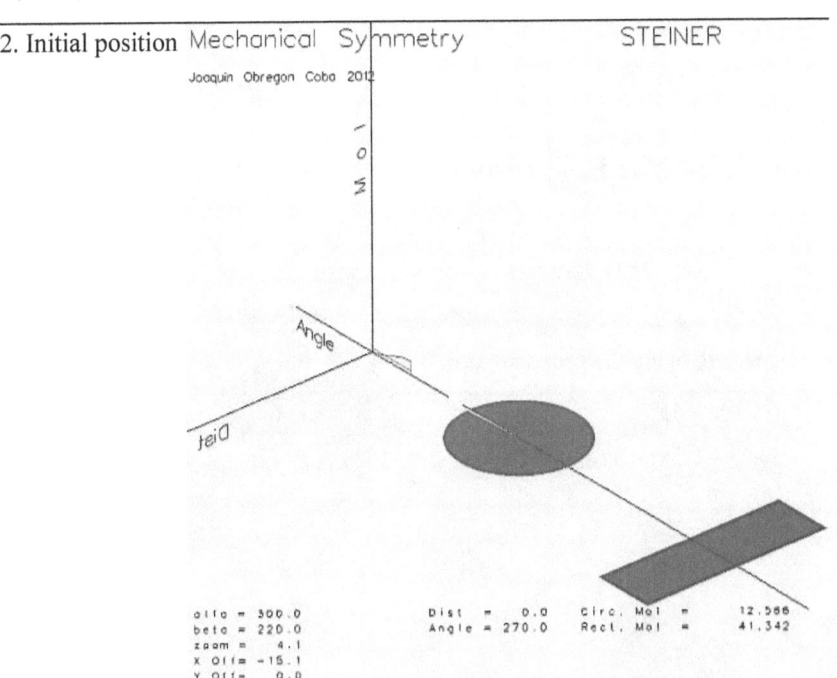

1. Circle

Circle with center on the axis.

```
(%i6)  Ic0:(%pi*r^4)/4;
```
$$(\%o6) 4\pi$$

```
(%i7)  float(%), numer;
```
$(\%o7) 12.56637061435917$

2. Rectangle

Rectangle with center of mass on the axis and longest side at 90° to the axis:
```
(%i8)  IRx0:(b*h^3)/12;
```
$$(\%o8) \frac{4\pi^3}{3}$$

```
(%i10) float(%), numer;
```
$(\%o10) 41.34170224039976$

```
(%i11) IRy0:(b^3*h)/12;
```
$$(\%o11) \frac{4\pi}{3}$$

```
(%i13) float(%), numer;
```
$(\%o13) 4.188790204786391$

Now Ixy:

```
(%i14) first(x,y):=integrate(x*y, y,-h/2,h/2);
```
$$(\%o14) \mathrm{first}(x,y) := \int_{\frac{-h}{2}}^{\frac{h}{2}} x\, y\, dy$$

```
(%i15) IRxy0:integrate(first(x,y), x,-b/2,b/2);
```
$(\%o15) 0$

As expected, being the principal axis:

3. First rotate 45°

```
(%i16) alfa:45/180*%pi;
```
$$(\%o16) \frac{\pi}{4}$$

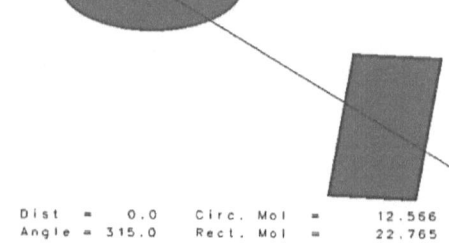

```
Dist  =    0.0      Circ. Mol =   12.566
Angle = 315.0       Rect. Mol =   22.765
```

1. Circle

Circle with the center on the axis.

(%i17) Ic1:(%pi*r^4)/4;

(%o17) 4π

(%i18) float(%), numer;

(%o18) 12.56637061435917

2. Rectangle rotation formulas:

(%i19) IRxnew(Ix,Iy,Ixy,a):=Ix*cos(a)^2+Iy*sin(a)^2-Ixy*sin(2*a);

(%o19) $\text{IRxnew}(Ix, Iy, Ixy, a) := Ix\cos(a)^2 + Iy\sin(a)^2 + (-Ixy)\sin(2a)$

(%i20) IRx1:IRxnew(IRx0,IRy0,IRxy0,alfa);

(%o20) $\dfrac{2\pi^3}{3} + \dfrac{2\pi}{3}$

(%i21) float(%), numer;

(%o21) 22.76524622259307

(%i22) IRynew(Ix,Iy,Ixy,a):=Ix*sin(a)^2+Iy*cos(a)^2+Ixy*sin(2*a);

(%o22) $\text{IRynew}(Ix, Iy, Ixy, a) := Ix\sin(a)^2 + Iy\cos(a)^2 + Ixy\sin(2a)$

(%i23) IRy1:IRynew(IRx0,IRy0,IRxy0,alfa);

(%o23) $\dfrac{2\pi^3}{3} + \dfrac{2\pi}{3}$

(%i24) float(%), numer;

(%o24) 22.76524622259307

(%i25) IRxynew(Ix,Iy,Ixy,a):=Ixy*cos(2*a)+(Ix+Iy)/2*sin(2*a);

(%o25) $\text{IRxynew}(Ix, Iy, Ixy, a) := Ixy\cos(2a) + \dfrac{Ix + Iy}{2}\sin(2a)$

(%i26) IRxy1:IRxynew(IRx0,IRy0,IRxy0,alfa);

(%o26) $\dfrac{\dfrac{4\pi^3}{3} + \dfrac{4\pi}{3}}{2}$

(%i27) float(%), numer;

(%o27) 22.76524622259307

4. Translation of 4 units:

Steiner

```
(%i28)  d:4;
```
(%o28)4

1. Circle

```
(%i29)  Ic2:Ic1+Ac*d^2;
```
(%o29)68π

```
(%i30)  float(%), numer;
```
(%o30)213.628300444106

2. Rectangle
```
(%i31)  IRx2:IRx1+Ar*d^2;
```
(%o31)$\dfrac{2\pi^3}{3}+\dfrac{194\pi}{3}$

```
(%i32)  float(%), numer;
```
(%o32)223.8271760523398

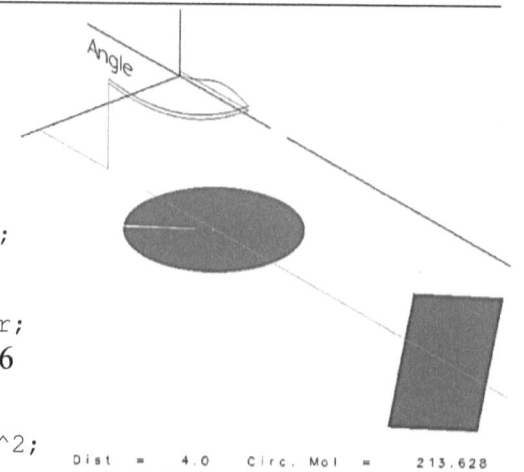

Dist = 4.0 Circ. MoI = 213.628
Angle = 315.0 Rect. MoI = 223.827

5. Translation of 1 more unit:

Steiner
```
(%i33)  d1:1;
```
(%o33)1

1. Circle
```
(%i34)  Ic3:Ic2+Ac*d1^2;
```
(%o34)72π

```
(%i35)  float(%), numer;
```
(%o35)226.1946710584651

----- No No No ----- Incorrect

2. Rectangle
```
(%i36)  IRx3:IRx2+Ar*d1^2;
```
(%o36)$\dfrac{2\pi^3}{3}+\dfrac{206\pi}{3}$

```
(%i37)  float(%), numer;
```
(%o37)236.393546666699 ----- NO NO NO ----- INCORRECTO

Let's do it right!

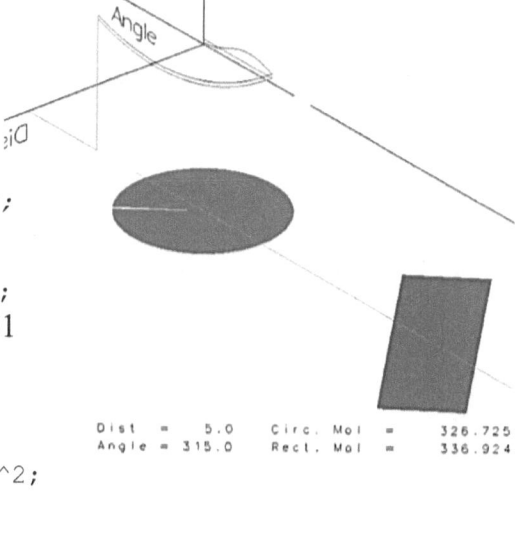

Dist = 5.0 Circ. MoI = 326.725
Angle = 315.0 Rect. MoI = 336.924

Mechanical Symmetry

3. Circle

(%i38) Ic3:Ic1+Ac*(d+d1)^2;

(%o38) 104π

(%i39) float(%), numer;

(%o39) 326.7256359733385

4. Rectangle

(%i40) IRx3:IRx1+Ar*(d+d1)^2;

(%o40) $\dfrac{2\pi^3}{3}+\dfrac{302\pi}{3}$

(%i41) float(%), numer;

(%o41) 336.9245115815724

And now with a different approach based on MoI definition, without using Steiner's theorem.

We define a circle and a rectangle with the same area.

(%i1) r:2;

(%o1) 2

---

1. Origin

First we calculate the MoI for a vertical infinitesimal.

(%i2) da:integrate(y^2, y,-y,y);

(%o2) $\dfrac{2y^3}{3}$

Circle equation:

(%i3) y:sqrt(r^2-x^2);

(%o3) $\sqrt{4-x^2}$

Integrate Fint*dx:

(%i4) Fint:ev(da, nouns);

(%o4) $\dfrac{2(4-x^2)^{\frac{3}{2}}}{3}$

(%i5) I:integrate(Fint, x,-r,r);

(%o5) 4π

```
(%i6) float(%), numer;
```
(%o6) 12.56637061435917

---

2. Moving 4 units:

```
(%i7) kill(y);
```
(%o7) *done*

```
(%i8) d:4;
```
(%o8) 4

First we calculate the MoI for a vertical infinitesimal.

```
(%i9) da:integrate(y^2, y,-y+d,y+d);
```
$$(\%o9)\frac{y^3+12y^2+48y+64}{3}+\frac{y^3-12y^2+48y-64}{3}$$

Circle equation:

```
(%i10) y:sqrt(r^2-x^2);
```
$$(\%o10)\sqrt{4-x^2}$$

Integrate Fint*dx:

```
(%i11) Fint:ev(da, nouns);
```
$$(\%o11)\frac{(4-x^2)^{\frac{3}{2}}+48\sqrt{4-x^2}+12(4-x^2)+64}{3}+$$
$$\frac{(4-x^2)^{\frac{3}{2}}+48\sqrt{4-x^2}-12(4-x^2)-64}{3}$$

```
(%i12) I:integrate(Fint, x,-r,r);
```
$$(\%o12)\,68\pi$$

```
(%i13) float(%), numer;
```
(%o13) 213.628300444106

---

3. Moving 1 more unit:

```
(%i14) kill(y);
```
(%o14) *done*

```
(%i15) d1:1;
```
(%o15) 1

Mechanical Symmetry

First we calculate the MoI for a vertical infinitesimal
```
(%i16) da:integrate(y^2, y,-y+d+d1,y+d+d1);
```
$$(\%o16)\frac{y^3+15y^2+75y+125}{3}+\frac{y^3-15y^2+75y-125}{3}$$

Circle equation:
```
(%i17) y:sqrt(r^2-x^2);
```
$$(\%o17)\sqrt{4-x^2}$$

Integra Fint*dx:
```
(%i18) Fint:ev(da, nouns);
```
$$(\%o18)\frac{(4-x^2)^{\frac{3}{2}}+75\sqrt{4-x^2}+15(4-x^2)+125}{3}+$$
$$\frac{(4-x^2)^{\frac{3}{2}}+75\sqrt{4-x^2}-15(4-x^2)-125}{3}$$

```
(%i19) I:integrate(Fint, x,-r,r);
```
$$(\%o19)104\pi$$

```
(%i20) float(%), numer;
```
$$(\%o20)326.7256359733385$$

Mechanical Symmetry

Mechanical Symmetry 3

New Answers

Mechanical Symmetry

3 New Answers — Mechanical Symmetry

We have an answer to our primary question: find the way to obtain sections with a constant MoI. From now on we will call those sections mechanically symmetric, or MS. So, our target now is finding MS sections.

3.1 Previous Test

Before starting with some new answers, let's review what we have been taught to date.

The test we are proposing consists of choosing the shape with a bigger MoI from several options, specifically between couples with 1, 2, 3, 4, 5 and 6 particles (or small circles). There is no geometry data because it is a concept test; it's not intended to make calculations and comparing numbers. You only need to know that the distance between particles is the same for each n.

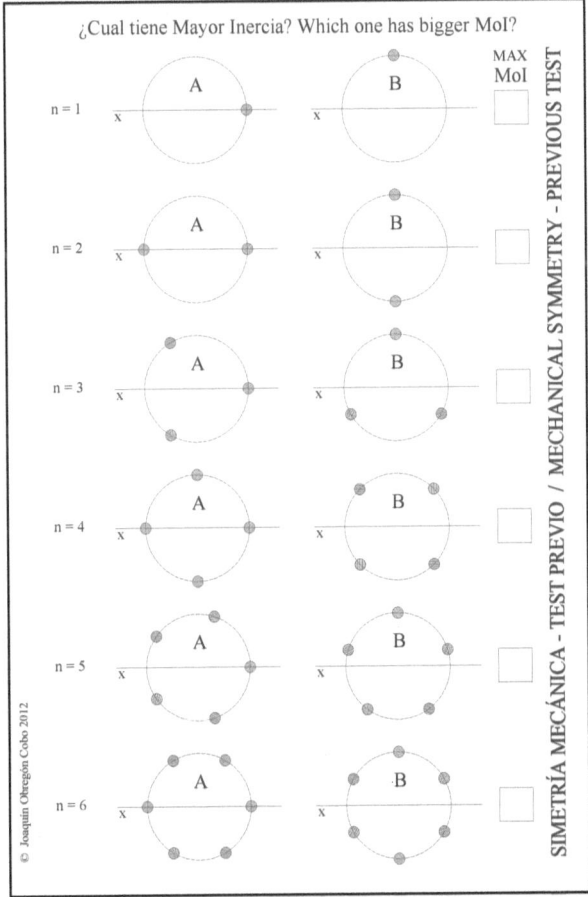

Let us guess your answers:

n = 1 → B , n = 2 → B , n = 3 → B , n = 4 → B , n = 5 → B , n = 6 → B

From 1 to 3 it is clear, and 4 seems clear, but 5 and 6 are not so clear because it looks like the value is similar.

The correct answers are B for 1 and 2, but for 3 onward, both values are identical.

What has happened?

When we want to increase the MoI for a section or shape, we try to put the strong material, or area for the shape, far from the axis from where we calculate MoI. This is perfectly correct and also smart, but as a side effect, it leads to us to identify a longer distance from the axis with a bigger value for MoI.

In the following chapter we will see the reasons explaining these surprising results.

3.2 First Approach

Start with the easy-to-solve before facing the complex; We are going to search for a section composed by particles. Why particles? They have a constant MoI. We can agree that a small circle could be considered a particle for our exposition, as far as circular shape has constant MoI.

3.2.1 Mechanically Symmetric Particles Sections

How to arrange k particles to make them MS? We have seen that only rotational symmetry shows some relationship with rotations (not very surprising). We locate the k particles, define a section with k-order symmetry, and check if MoI is constant or not. The first step is calculating the value of the MoI. The formula used for doing it is:

$$I_k = kI_n + a \sum_{n=1}^{k} d_n^2 \qquad [2]$$

Eq. 8 Particles MoI – First Summation

Because a and kI_n are constants, the only change can come from:

$$\sum_{n=1}^{k} d_n^2 = \sum_{n=1}^{k} (r sen\alpha_n)^2 = r^2 \sum_{n=1}^{k} (sen\alpha_n)^2$$

Eq. 9 Particles MoI – Variable Summand

We calculate this for different values of $\alpha_n = \frac{2\pi}{k} n + \alpha_0$ as a function of α_0 and k obtaining a value of $\frac{k}{2}$ for every k greater than 2, with no dependence on α_0.

This means that the MoI does not depend on the orientation of the section for $k \geq 3$.

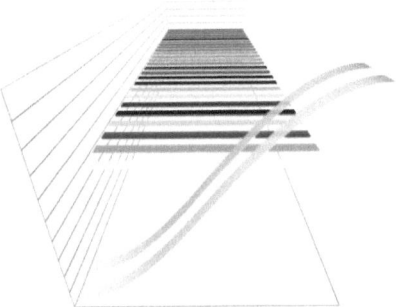

Fig. 19 Particles MoI - Summations

In the graph we can see in the vertical axis $y = \dfrac{\sum_{n=1}^{k} \sin^2\left(\dfrac{2\pi}{k} n + \alpha_0\right)}{k}$ where α_0 goes from 0° to 90° from left to right, and k goes from 1 to 28 to the back, showing constant value for $k \geq 3$.

Using a computer program, we made the sums again for double checking. You can find it complete in appendix 2.

```
REM LOOP FOR k from 1 TO 16
REM     LOOP FOR orientation from 0 to 90 degrees
REM         LOOP TO sum every particle (i from 1 TO k)
FOR k=1.0 TO SAMPLES STEP 1.0
    PRINT USING " ##   ": k;
    REM ang is the angle between particles
    LET ang = PI * 2.0 / k
    REM angIni defines the rotation of the section as the
    REM         initial angle for the first particle
    FOR angIni = 0.0 TO PI/2.0 STEP ANGININCRAD
        LET sum = 0.0
        FOR i=1 TO k
            REM alfa is the angle for every particle
            LET alfa = ang * i + angIni
            LET sum = sum + SIN(alfa)*SIN(alfa)
        NEXT i
        REM Now sum has the sum of sin2
        LET sum = sum / k
        REM Now it contains the constant sum/k (for k≥ 3)
        REM And we print it
        PRINT USING "----%.###": sum;
    NEXT angIni
    PRINT
NEXT k
```

Now we know that the simplified formula for calculating the moment for a set of k round elements with k-order symmetry is:

[a]
$$I_k = \frac{k \cdot a \cdot r^2}{2} \quad \forall\, k \in \mathbb{N}, k \geq 3$$

Eq. 10 Particles MoI – Approximate Formula

If we prefer an exact calculation, including each particle's MoI (I_n):

[b]
$$I_k = kI_n + a\sum_{n=1}^{k} d_n^2 = k\left(I_n + \frac{a \cdot r^2}{2}\right)$$

$$\forall\, k \in \mathbb{N}, k \geq 3$$

Eq. 11 Particles MoI – Exact Formula

Being

k Number of particles

a Area for each particle

r Radius of the circumference

I_n Moment of Inertia of each particle

Mechanical Symmetry

At this point we have yet an answer: we know how to obtain a mechanically symmetric particles section: arranging particles with *k*-order rotational symmetry.

Here is the mathematical proof.

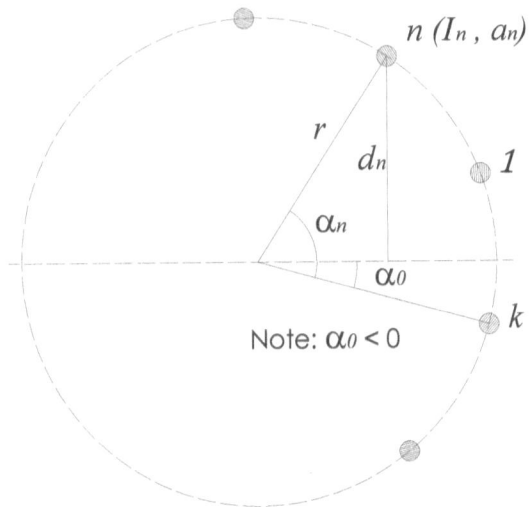

Fig. 20 Particles MoI - Proof

n^{th} Particle angle:

$$\alpha_n = \frac{2\pi}{k} n + \alpha_0 \qquad \forall\, n \in \mathbb{N}, n \le k$$

Distance to axis:

$$d_n = r \cdot \sin\alpha_n$$

Moment and area for the nth particle:

$$I_n \qquad a_n$$

MoI for the *k* particles (Steiner)

$$I_k = kI_n + \sum_{n=1}^{k} a_n d_n^2 \qquad [1]$$

Identical items with constant MoI (Otherwise no rotational symmetry):

Mechanical Symmetry

[2] $$I_k = kI_n + a\sum_{n=1}^{k} d_n^2$$

Where

[3] $$\sum_{n=1}^{k} d_n^2 = \sum_{n=1}^{k} (r\, sen\,\alpha_n)^2 = \sum_{n=1}^{k} \frac{r^2}{2}(1-\cos 2\alpha_n) = \frac{r^2}{2}\left(\sum_{n=1}^{k} 1 - \sum_{n=1}^{k} \cos 2\alpha_n\right)$$

For the calculation of the second summand, we use Euler's notation.

$$e^{\alpha i} = \cos\alpha + i\,\sin\alpha$$

Eq. 12 Euler's notation

$$\sum_{n=1}^{k} \cos 2\alpha_n + i\sum_{n=1}^{k} \sin 2\alpha_n = \sum_{n=1}^{k} e^{i2\alpha_n} = \sum_{n=1}^{k} e^{i2\left(\frac{2\pi}{k}n+\alpha_0\right)} =$$

$$= \sum_{n=1}^{k} e^{i\frac{4\pi}{k}n + 2\alpha_0} = \sum_{n=1}^{k} e^{i\frac{4\pi}{k}n} e^{i2\alpha_0}$$

This is the sum of a geometric progression from 1 to k

$$a_n = e^{i2\alpha_n} = e^{i2\alpha_0} e^{i\frac{4\pi}{k}n} \quad \overset{a_{n+1}=a_n p}{\Longrightarrow} \quad p = e^{i\frac{4\pi}{k}}$$

$$a_1 = e^{i2\alpha_0} e^{i\frac{4\pi}{k}}$$

$$a_k = e^{i2\alpha_0} e^{i\frac{4\pi}{k}k} = e^{i2\alpha_0} e^{i4\pi}$$

Formula for the sum:

$$S_k = \frac{a_1 - p\cdot a_k}{1-p}$$

Replacing

$$S_k = \frac{e^{i2\alpha_0} e^{i\frac{4\pi}{k}} - e^{i2\alpha_0} e^{i\frac{4\pi}{k}} \cdot e^{i4\pi}}{1 - e^{i\frac{4\pi}{k}}} = \frac{\left(1 - e^{i4\pi}\right)\cdot e^{i2\alpha_0} e^{i\frac{4\pi}{k}}}{1 - e^{i\frac{4\pi}{k}}} = \frac{0}{1 - e^{i\frac{4\pi}{k}}}$$

Mechanical Symmetry

Evaluating the denominator for the values of k:

$$k = 1$$

$$e^{i\frac{4\pi}{k}} = e^{i4\pi} = \cos 4\pi + i\sin 4\pi = 1$$

$$1 - e^{i\frac{4\pi}{k}} = 1 - 1 = 0$$

$$S_k \neq 0$$

$$k = 2$$

$$e^{i\frac{4\pi}{k}} = e^{i2\pi} = \cos 2\pi + i\sin 2\pi = 1$$

$$1 - e^{i\frac{4\pi}{k}} = 1 - 1 = 0$$

$$S_k \neq 0$$

$$k = 3$$

$$e^{i\frac{4\pi}{k}} = e^{i\frac{4}{3}\pi} = \cos\tfrac{4}{3}\pi + i\sin\tfrac{4}{3}\pi = -0'5 - 0'866\,i$$

$$1 - e^{i\frac{4\pi}{k}} \neq 0$$

$$S_k = 0$$

$$k = 4$$

$$e^{i\frac{4\pi}{k}} = e^{i\pi} = \cos \pi + i\sin \pi = -1$$

$$1 - e^{i\frac{4\pi}{k}} = 1 + 1 \neq 0$$

$$S_k = 0$$

$$k \geq 5$$

$$e^{i\frac{4\pi}{k}} = e^{iA\pi} \quad 0 \leq A \leq 1 \Rightarrow \cos A\pi + i\sin A\pi \neq 1$$

$$1 - e^{i\frac{4\pi}{k}} \neq 0$$

$$S_k = 0$$

We arrive to

$$S_k = 0 \quad \forall k \in \mathbb{N}, k \geq 3$$

Eq. 13 MoI Particles – Null Sum Proof

And then

$$S_k = \sum_{n=1}^{k} e^{i\frac{4\pi}{k}n} e^{i2\alpha_0} = \sum_{n=1}^{k} \cos 2\alpha_n + i\sum_{n=1}^{k} \sin 2\alpha_n = 0 + 0i \Rightarrow$$

$$\Rightarrow \sum_{n=1}^{k} \cos 2\alpha_n = 0$$

Replacing in [3]:

[4] $$\sum_{n=1}^{k} d_n^2 = \frac{r^2}{2}\left(\sum_{n=1}^{k} 1 - \sum_{n=1}^{k} \cos 2\alpha_n\right) = \frac{r^2}{2}\left(\sum_{n=1}^{k} 1 - 0\right) = \frac{k \cdot r^2}{2}$$

And finally going back to [2]:

[5] ≡ [b] $$I_k = kI_n + a\sum_{n=1}^{k} d_n^2 = kI_n + a\frac{k \cdot r^2}{2} = k\left(I_n + \frac{a \cdot r^2}{2}\right)$$

$$\forall k \in \mathbb{N}, k \geq 3$$

Eq. 14 Particles MoI – Exact Formula Proof

We see that the formula obtained has no dependence on α_0, meaning that MoI is constant for any orientation for $k \geq 3$. Finally we can assert that a set of k particles located with k-order rotational symmetry has constant MoI —they have mechanical symmetry— for $k \geq 3$.

3.2.2 Formulas for Mechanically Symmetric Particles Sections

As a result from the previous section, we have arrived to prove that for a set of particles located with k-order rotational symmetry:

$$I_k = k\left(I_n + \frac{a \cdot r^2}{2}\right) \quad \forall\, k \in \mathbb{N}, k \geq 3 \qquad [b] \equiv [5]$$

Eq. 15 Mol Particles - Exact

Being
k Number of particles
a Area for each particle
r Radius of the circumference where particles lay in
I_n Moment for each particle

Sometimes kI_n is irrelevant compared to $\dfrac{k \cdot a \cdot r^2}{2}$ so we can use:

$$I_k = \frac{k \cdot a \cdot r^2}{2} \quad \forall\, k \in \mathbb{N}, k \geq 3 \qquad [a] \equiv [6]$$

Eq. 16 Mol Particles - Approximate

With an accuracy

$$\Delta I_k = \frac{k\left(I_n + \frac{a \cdot r^2}{2}\right) - \frac{k \cdot a \cdot r^2}{2}}{\frac{k \cdot a \cdot r^2}{2}} = \frac{\left(I_n + \frac{a \cdot r^2}{2}\right) - \frac{a \cdot r^2}{2}}{\frac{a \cdot r^2}{2}} = \frac{2I_n}{a \cdot r^2} \qquad [7]$$

Eq. 17 Mol Particles Accuracy

Regarding the accuracy of the formula [6] as of formula [7], we see that there is no dependency of the number of items k.

We can see in *Table 1* some examples. The values used in the table are typical in concrete steel reinforcement, rebars of a reinforced concrete pile or column.

Say that the diameter of the circle corresponds with the concrete column diameter, and the particles diameter is the rebar diameter (no cover to simplify). This means that calculations made with the simplified formula [6] are accurate enough for any reinforced concrete calculation.

Table 1 Particles MoI – Approximate formula accuracy

Circle Diameter (cm)	Number of Particles	Particles Diameter (mm)	Exact MoI (cm^4)	Approx. MoI (cm^4)	Accuracy (‰)
60	8	16	7240,8	7238,2	**0,36**
60	8	25	17686,8	17671,5	**0,87**
60	8	40	45339,5	45238,9	**2,22**
60	16	16	14481,6	14476,5	**0,36**
60	16	25	35373,6	35342,9	**0,87**
60	16	40	90678,9	90477,9	**2,22**
60	24	16	21722,4	21714,7	**0,36**
60	24	25	53060,4	53014,4	**0,87**
60	24	40	136018,4	135716,8	**2,22**
120	8	16	28955,5	28952,9	**0,09**
120	8	25	70701,2	70685,8	**0,22**
120	8	40	181056,3	180955,7	**0,56**
120	16	16	57911,0	57905,8	**0,09**
120	16	25	141402,3	141371,7	**0,22**
120	16	40	362112,5	361911,5	**0,56**
120	24	16	86866,5	86858,8	**0,09**
120	24	25	212103,5	212057,5	**0,22**
120	24	40	543168,8	542867,2	**0,56**
200	8	16	80427,3	80424,8	**0,03**
200	8	25	196364,9	196349,5	**0,08**
200	8	40	502755,4	502654,8	**0,20**
200	16	16	160854,7	160849,5	**0,03**
200	16	25	392729,8	392699,1	**0,08**
200	16	40	1005510,7	1005309,6	**0,20**
200	24	16	241282,0	241274,3	**0,03**
200	24	25	589094,6	589048,6	**0,08**
200	24	40	1508266,1	1507964,5	**0,20**

Note that accuracy is not % but ‰.

You can find further detail about this subject in appendix 2.

3.3 Generalization

Now we know particles systems that have mechanical symmetry, and we also know their formulas. This is useful for steel reinforcement sections; it has its own value by itself, but we want to extend the concept to any section.

3.3.1 Previous Step

We start with a favorable case: a section with rotational symmetry able to be decomposed in small elements far from the center, to get a small ΔI_k (see [7]). We use a tube section with a small thickness e and radius r much bigger than the thickness.

Dividing the section in k elements as illustrated in the figure and using [6]:

$$I_k = \frac{k \cdot a \cdot r^2}{2} \quad \forall k \in N : k \geq 3 \quad [6]$$

$$a = \frac{2\pi r}{k} e$$

$$I_k = \frac{k \cdot \frac{2\pi r}{k} e \cdot r^2}{2}$$

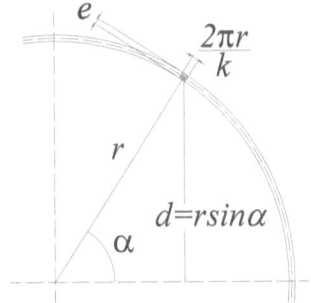

Fig. 21 Thin Wall Tube MoI – MS Formula

$$I = \pi e r^3 \quad [T1]$$

Eq. 18 Thin Wall Tube MoI – MS Formula

Using MoI definition:

$$d = \rho \cdot \sin\theta$$

$$d\Omega = \rho \cdot d\theta \cdot e$$

$$I = \int_\Omega d^2 d\Omega = \int_0^{2\pi} \rho^2 \sin^2\theta \cdot \rho \cdot d\theta \cdot e =$$

$$= e\rho^3 \int_0^{2\pi} \sin^2\theta \, d\theta$$

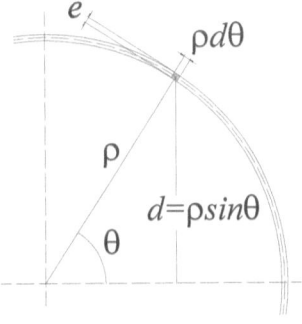

Fig. 22 Thin Wall Tube MoI

$$I = \pi e \rho^3 \quad [T2]$$

Eq. 19 Thin Wall Tube MoI

We see that [T1]=[T2], and therefore our approximate formula is valid.

3.3.2 Generic Section

Now we try to generalize. To do so, we get an arbitrary section, with the only condition that it has k-order rotational symmetry.

From [1] we see that the moment of the elements changes for every different α, then being I_u and I_v principal moments in the center of mass (CoM):

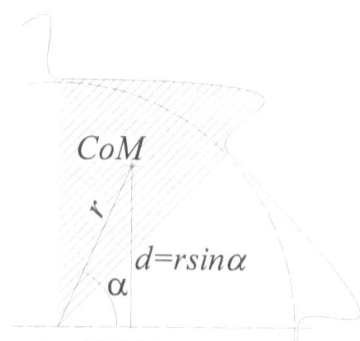

Fig. 23 MoI Generic Section

$$I = \int_\Omega d^2 d\Omega = \sum_{n=1}^{k} \int_{\Omega_n} d^2 d\Omega = \sum_{n=1}^{k}\left(I_n + a_n d_n^2\right) = I_k$$

Being

$$I_n = I_u \cos^2\alpha_n + I_v \operatorname{sen}^2\alpha_n \quad ; \quad \alpha_n = \tfrac{2\pi}{k} n + \alpha_0 \qquad \forall n \in \mathbb{N}, n \le k$$

We get

$$I = \sum_{n=1}^{k}\left(I_n + a_n d_n^2\right) = \sum_{n=1}^{k}\left(I_u \cos^2\alpha_n + I_v \sin^2\alpha_n + a_n d_n^2\right) =$$

$$= \sum_{n=1}^{k} I_u \cos^2\alpha_n + \sum_{n=1}^{k} I_v \sin^2\alpha_n + \sum_{n=1}^{k} a_n d_n^2$$

I_u, I_v and $a_n = a$ are constants $\Rightarrow$

$$\Rightarrow I = I_u \sum_{n=1}^{k}\cos^2\alpha_n + I_v \sum_{n=1}^{k}\sin^2\alpha_n + a \sum_{n=1}^{k} d_n^2$$

As far as

$$\left. \begin{array}{l} \cos^2\alpha + \sin^2\alpha = 1 \\[4pt] \displaystyle\sum_{n=1}^{k}\sin^2\alpha_n = \frac{k}{2} \quad \forall k \in \mathbb{N}: k \ge 3 \end{array} \right\} \Rightarrow \sum_{n=1}^{k}\cos^2\alpha_n = \frac{k}{2} \quad \forall k \in \mathbb{N}: k \ge 3$$

Mechanical Symmetry

We arrive at

$$I_k = I_u \frac{k}{2} + I_v \frac{k}{2} + ar^2 \frac{k}{2}$$

And finally at a formula for any section with mechanical symmetry.

$$I = \frac{k}{2}\left(I_u + I_v + ar^2\right) \quad \forall k \in \mathbb{N} : k \geq 3 \qquad [d]$$

Eq. 20 MoI Generic Section with Mechanical Symmetry.

Being
k Number of elements
a Area for each element
r Radius of the circumference
I_u, I_v Principal MoI for each element (at center of mass)

Obviously, the approximate formula and accuracy are:

$$I_k = \frac{k \cdot a \cdot r^2}{2} \quad \forall \, k \in \mathbb{N}, k \geq 3 \qquad [a] \equiv [6] \equiv [e]$$

Eq. 21 MoI Generic Section with MS – Approximate

$$\Delta I_k = \frac{k\left(I_u + I_v + \dfrac{a \cdot r^2}{2}\right) - \dfrac{k \cdot a \cdot r^2}{2}}{\dfrac{k \cdot a \cdot r^2}{2}} = 2\frac{I_u + I_v}{a \cdot r^2} \qquad [f]$$

Eq. 22 MoI Generic Section with MS – Accuracy

3.E Exercises about Mechanical Symmetry

As an exercise and additional verification, we now check whether the I_{ky} perpendicular to $I_k = I_x = I_{kx}$ is the same as them. Then we will have increased certainty about I_k being constant, as the absence of the angle in the formula indicates.

$$I_{ky} = \int_\Omega d^2 d\Omega = \sum_{n=1}^{k} \int_{\Omega_n} d^2 d\Omega = \sum_{n=1}^{k}\left(I_n + a_n d_n^2\right)$$

$$I_n = I_y = I_u \sin^2 \alpha_n + I_v \cos^2 \alpha_n \;;\; \alpha_n = \frac{2\pi}{k} n + \alpha_0 \qquad \forall n \in \mathbb{N}, n \leq k$$

$$I_{ky} = \sum_{n=1}^{k}\left(I_n + a_n d_n^2\right) = \sum_{n=1}^{k}\left(I_u \sin^2 \alpha_n + I_v \cos^2 \alpha_n + a_n d_n^2\right)$$

Mechanical Symmetry

$$I_{ky} = I_u \frac{k}{2} + I_v \frac{k}{2} + ar^2 \frac{k}{2} \equiv I_k$$

We arrive to the same formula, verifying that the moment I_k is constant, we can state that the second moment of area ellipse is a circle, or that any coordinated system passing by the center is principal, so I_{xy} is zero.

To finish and make a final check, we calculate the value of the characteristics for a circular section with our formulas, as composed by four quarters of circle, and we compare them with the known value.

Table 2 Known Values for the Circle

Center of Mass	$(0,0)$
Area	πr^2
Moments	$I_{xy} = 0 \quad I_k = I_x = I_y = I_u = I_v = \dfrac{\pi r^4}{4}$
Radius of Gyration	$i_k = i_x = i_y = i_u = i_v = \dfrac{r}{2}$
Polar Moment	$I_P = \dfrac{\pi r^4}{2}$

Table 3 Known Values for Quarter of a Circle

Center of Mass	$\left(\dfrac{4r}{3\pi}, \dfrac{4r}{3\pi}\right)$
Area	$\dfrac{\pi r^2}{4}$
Moments	$I_{xy} = 0 \quad I_k = I_x = I_y = I_u = I_v = \dfrac{\pi r^4}{16}$
Radius of Gyration	$i_k = i_x = i_y = i_u = i_v = \dfrac{r}{2}$
Polar Moment	$I_P = \dfrac{\pi r^4}{8}$

Mechanical Symmetry

Say r_c is the radius of the quarter of a circle; we calculate the radius r from the center of symmetry to the center of mass of the quarter:

$$r = \sqrt{\left(\frac{4r_c}{3\pi}\right)^2 + \left(\frac{4r_c}{3\pi}\right)^2} = \frac{4r_c}{3\pi}\sqrt{2}$$

Area for the quarter and number of elements:

$$a = \frac{\pi r_c^2}{4} \quad k = 4$$

Using Steiner's theorem, we get I_u and I_v at the center of masses.

$$I_u = I_v = \frac{\pi r_c^4}{16} - \frac{\pi r_c^2}{4}\left(\frac{4r_c}{3\pi}\right)^2 = \frac{\pi r_c^4}{16} - \frac{4r_c^4}{9\pi} = r_c^4\left(\frac{\pi}{16} - \frac{4}{9\pi}\right)$$

Substituting in [d]:

$$I_k = \frac{k}{2}(I_u + I_v + ar^2) =$$

$$= \frac{4}{2}\left[r_c^4\left(\frac{\pi}{16} - \frac{4}{9\pi}\right) + r_c^4\left(\frac{\pi}{16} - \frac{4}{9\pi}\right) + \frac{\pi r_c^2}{4}\left(\frac{4r_c}{3\pi}\sqrt{2}\right)^2\right]$$

$$I_k = 2\left(\frac{\pi r_c^4}{8} - \frac{8r_c^4}{9\pi} + \frac{8r_c^4}{9\pi}\right) = \frac{\pi r_c^4}{4}$$

Exactly the same.

Mechanical Symmetry

Mechanical Symmetry

3.4 Definition and Theorem

A section has mechanical symmetry if the moment of inertia (second moment of area) related to an axis through its center of mass remains constant when the section is rotated around its center of mass.

It is sufficient condition to state that a section has mechanical symmetry having *k*-order rotational symmetry, if and only if *k* is an integer greater or equal than 3.

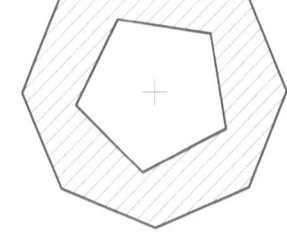

3.5 Corollary

If a section is composed by figures with mechanical symmetry, it will have mechanical symmetry even if it does not have rotational symmetry.

Fig. 24 Mechanical Symmetry without Rotational Symmetry

3.6 Necessity and Sufficiency

We have seen that the rotational symmetry is a sufficient condition for mechanical symmetry. This raises the question: Is it necessary? Or does the fact that a section have mechanical symmetry imply that it has rotational symmetry? And if so, is it necessary that the order of symmetry is equal to or larger than 3?

The immediately preceding corollary answers this question. As illustrated in *Fig. 24 Mechanical Symmetry without Rotational Symmetry*, sections without rotational symmetry can have mechanical symmetry; the outer perimeter of the section is an octagon, and a pentagon is inside, sharing the center of symmetry, so that there is no geometric symmetry, but the MoI of the section is constant because it is the difference between the MoI of the octagon and pentagon, which are constant. Clearly, sections composed by several elements do not need to have rotational symmetry to have mechanical symmetry.

$$I_{Total} = I_{Octagon} - I_{Pentagon} = Ct_1 - Ct_2 = \text{Constant}$$

But we still need to answer the question for simple sections, defining them as the area encircled by a single line.

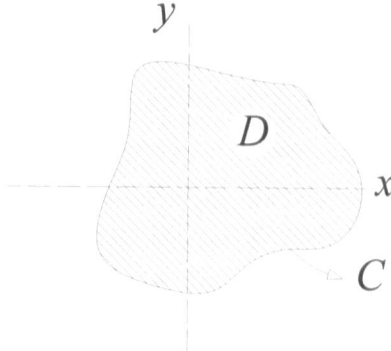

Then we start from: a line with an area inside it, and the constant value of the MoI for this area when rotating around the center of mass.

We use the fact that $I_{xy}=0$ for every angle to assess the constant value of the MoI.

$$I_{xy} = \iint_D xy\,d\Omega = \iint_D xy\,dx\,dy = 0$$

To face the problem, we use Green's theorem; it is explained in detail in the appendix *Ap. 4.2*.

$$I_{xy} = \iint_D xy\,d\Omega = \iint_D xy\,dx\,dy$$

$$\left(\frac{\partial M}{\partial x} - \frac{\partial L}{\partial y}\right) = xy \Rightarrow \begin{cases} M = 0 & L = -x\dfrac{y^2}{2} \\ M = \dfrac{x^2}{2}y & L = 0 \end{cases}$$

$$\oint_C (L\,dx + M\,dy) = \begin{cases} [G1] & \oint_C -x\dfrac{y^2}{2}\,dx \\ [G2] & \oint_C \dfrac{x^2}{2}y\,dy \end{cases}$$

Now we can calculate the value of the integral over the area D of the section using the line integral over the perimeter C. We are going to use polar coordinates with the center of mass of the section as pole or origin for the reference system.

$$\begin{cases} x = \rho\cos\theta & dx = -\rho\sin\theta\,d\theta \\ y = \rho\sin\theta & dy = \rho\cos\theta\,d\theta \end{cases}$$

First we integrate [G1]:

$$I_{xy} = \oint_C -x\frac{y^2}{2}dx = \oint_C -\rho\cos\theta\frac{\rho^2\sin^2\theta}{2}(-\rho\sin\theta d\theta) =$$

$$= \oint_C \frac{\rho^4\cos\theta\sin^3\theta}{2}d\theta = \frac{1}{4}\oint_C \rho^4\sin^2\theta\sin 2\theta\, d\theta = 0$$

$$\left.\begin{array}{l} I_{xy} = \dfrac{1}{4}\oint_C \rho^4\sin 2\theta\,(1-\cos^2\theta)d\theta = \\ \\ = \dfrac{1}{4}\left[\oint_C \rho^4\sin 2\theta\, d\theta - \oint_C \rho^4\sin 2\theta\cos^2\theta d\theta\right] = 0 \\ \\ \text{From G2: } \oint_C \rho^4\sin 2\theta\cos^2\theta d\theta = 0 \end{array}\right\} \Rightarrow \oint_C \rho^4\sin 2\theta\, d\theta = 0$$

We integrate [G2], too:

$$I_{xy} = \oint_C y\frac{x^2}{2}dx = \oint_C \rho\sin\theta\frac{\rho^2\cos^2\theta}{2}\rho\cos\theta d\theta =$$

$$= \oint_C \frac{\rho^4\sin\theta\cos^3\theta}{2}d\theta = \frac{1}{4}\oint_C \rho^4\cos^2\theta\sin 2\theta\, d\theta = 0$$

$$\left.\begin{array}{l} I_{xy} = \dfrac{1}{4}\oint_C \rho^4\sin 2\theta\,(1-\sin^2\theta)d\theta = \\ \\ = \dfrac{1}{4}\left[\oint_C \rho^4\sin 2\theta\, d\theta - \oint_C \rho^4\sin 2\theta\sin^2\theta d\theta\right] = 0 \\ \\ \text{From G1: } \oint_C \rho^4\sin 2\theta\sin^2\theta d\theta = 0 \end{array}\right\} \Rightarrow \oint_C \rho^4\sin 2\theta\, d\theta = 0$$

Both ways we conclude that the section having constant MoI implies that the following equation will be fulfilled:

Mechanical Symmetry

With θ_0 the angle of rotation of the section.

[g] $\qquad \oint_C \rho^4 \sin 2\theta \, d\theta = \oint_C \rho(\theta,\theta_0)^4 \sin 2\theta \, d\theta = 0 \quad \forall \theta_0 \in [-\pi,\pi]$

Eq. 23 Necessary Condition for Mechanical Symmetry

We have to remember that the mathematical formulation used restricts the problem to a closed, convex line. This perfectly fits, from a physical point of view, with the problem we face. The mathematical expression for this is:

[h] $\qquad \rho \equiv \rho(\theta,\theta_0) \begin{cases} \rho \geq 0 \\ \rho(\theta+2\pi,\theta_0) = \rho(\theta,\theta_0) \end{cases}$

Eq. 24 Additional Necessary Conditions for Mechanical Symmetry

If we analyze [g], we see that, for example, an even function like an ellipse with parameters a and b, and one horizontal axis fulfills the condition.

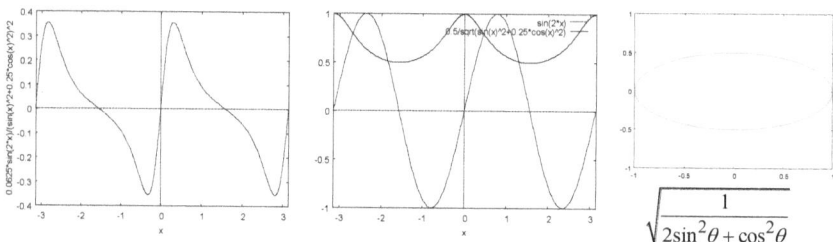

Fig. 25 Necessary Condition for Mechanical Symmetry – Even Function

$$\oint_C \rho^4 \operatorname{sen} 2\theta \, d\theta = \int_{-\pi}^{\pi} \left(\frac{1}{2\sin^2\theta + \cos^2\theta} \right)^4 \operatorname{sen} 2\theta \, d\theta = 0$$

But this is only when the axis is horizontal. If we rotate the section $\pi/3$ radians, it does not fulfill the condition.

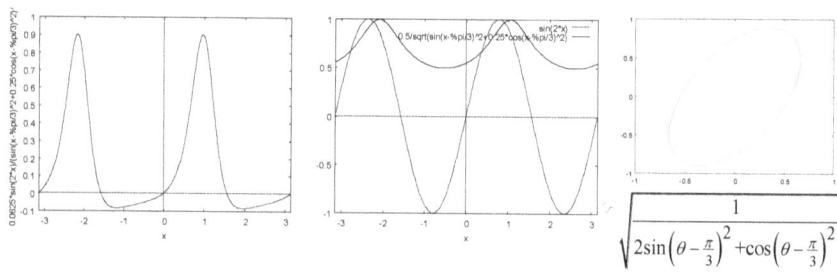

Fig. 26 Necessary Condition for Mechanical Symmetry – Rotated Even Function

Mechanical Symmetry

$$\rho'(\theta) = \rho(\theta - \tfrac{\pi}{3})$$

$$\int_{-\pi}^{\pi} \left(\frac{1}{2\sin^2\left(\theta - \tfrac{\pi}{3}\right) + \cos^2\left(\theta - \tfrac{\pi}{3}\right)} \right)^4 \sin 2\theta \, d\theta = 1.0 \neq 0$$

From the previous example, from the obvious one that is the circle ($\rho = R$ =constant), and from some others we see that being an even function is not enough, but to nullify equation [g], the function $\rho = \rho(\theta, \theta_0)$ needs to be periodic in a way that ρ raised to the forth power and multiplied by the sine of the double of θ is an odd function or any orientation. In other words, the key factor is the dependence of ρ on the rotation (what is not surprising). We know from [h] that the periodicity of the function must be a multiple or divisor of 2π, then:

$$\left. \begin{array}{l} \text{From [h] } \rho(\theta,\theta_0) = \rho(\theta + 2\pi, \theta_0) \\ \text{From [g] } \rho(\theta,\theta_0) = \rho(\theta + T, \theta_0) \end{array} \right\} \Rightarrow T = \frac{2\pi}{k} \; \forall k \in \mathbb{N}$$

Following Fourier's analysis, every periodic function with a period T can be represented as a sum of an infinite number of harmonic functions:

$$f(\theta) = \sum_{n=1}^{\infty} A_n \cdot \sin\left(\frac{2\pi n \theta}{T} + \theta_n\right)$$

Eq. 25 Fourier's Series

Applying Fourier's analysis to ρ^4 in our formula, we get the following (θ_0 is omitted now to simplify):

$$\rho^4(\theta) = \sum_{n=1}^{\infty} A_n \cdot \sin(kn\theta + \theta_n)$$

And going back to [g]:

$$\oint_C \rho^4 \sin 2\theta \, d\theta = \int_{-\pi}^{\pi} \rho^4(\theta) \sin 2\theta \, d\theta =$$

$$= \int_{-\pi}^{\pi} \left[\sum_{n=1}^{\infty} A_n \cdot \sin(kn\theta + \theta_n) \right] \sin 2\theta \, d\theta =$$

$$= \sum_{n=1}^{\infty} \left[\int_{-\pi}^{\pi} A_n \cdot \sin(kn\theta + \theta_n) \sin 2\theta \, d\theta \right] =$$

$$= \sum_{n=1}^{\infty}\left[A_n \int_{-\pi}^{\pi} \sin(kn\theta + \theta_n)\sin 2\theta \, d\theta\right] =$$

$$= \sum_{n=1}^{\infty}\left[A_n \int_{-\pi}^{\pi} (\sin kn\theta \cos\theta_n + \cos kn\theta \sin\theta_n)\sin 2\theta \, d\theta\right] =$$

$$= \sum_{n=1}^{\infty} A_n \left[\int_{-\pi}^{\pi} \sin kn\theta \cos\theta_n \sin 2\theta \, d\theta + \int_{-\pi}^{\pi} \cos kn\theta \sin\theta_n \sin 2\theta \, d\theta\right] = 0$$

The values of A_n and θ_n are constants, unknown and with values from which we cannot assume anything without distorting our model. When we group these constants, our equation is:

$$\begin{cases} A_n \cos\theta_n = B_n \\ A_n \sin\theta_n = C_n \\ A_n^2 = B_n^2 + C_n^2 \end{cases}$$

$$\sum_{n=1}^{\infty}\left[B_n \int_{-\pi}^{\pi} \text{sen } kn\theta \sin 2\theta \, d\theta + C_n \int_{-\pi}^{\pi} \cos kn\theta \sin 2\theta \, d\theta\right] = 0$$

We write back θ_0 in our equation:

[i]
$$\sum_{n=1}^{\infty}\left[B_n \int_{-\pi}^{\pi} \sin(kn\theta + \theta_0)\sin 2\theta \, d\theta + C_n \int_{-\pi}^{\pi} \cos(kn\theta + \theta_0)\sin 2\theta \, d\theta\right] = 0$$

$$\forall \theta_0 \in [-\pi, \pi]$$

Eq. 26 Necessary Condition for Mechanical Symmetry – Fourier

To understand what this equation is telling us, we divide it and analyze each component in detail. As previously said, B_n and C_n are constants, unknown, but they can be calculated from the original ρ function; they are the values "describing" the shape of the function. We conclude then that the key is in the integrals.

$$\int_{-\pi}^{\pi} \sin(kn\theta + \theta_0)\sin 2\theta \, d\theta \quad \text{and} \quad \int_{-\pi}^{\pi} \cos(kn\theta + \theta_0)\sin 2\theta \, d\theta$$

We simplify them with $\phi = kn$

$$factor_1 = \int_{-\pi}^{\pi} \sin(\phi\theta + \theta_0)\sin 2\theta \, d\theta$$

$$factor_2 = \int_{-\pi}^{\pi} \cos(\phi\theta + \theta_0)\sin 2\theta \, d\theta$$

Mechanical Symmetry

Being

$$\int \sin(\phi\theta + \theta_0) \sin 2\theta \, d\theta = \frac{\sin(\theta_0 + \phi\theta - 2\theta)}{2(\phi - 2)} - \frac{\sin(\theta_0 + \phi\theta + 2\theta)}{2(\phi + 2)}$$

$$\int \cos(\phi\theta + \theta_0) \sin 2\theta \, d\theta = \frac{\cos(\theta_0 + \phi\theta - 2\theta)}{2(\phi - 2)} - \frac{\cos(\theta_0 + \phi\theta + 2\theta)}{2(\phi + 2)}$$

We analyze now what happens for different values of ϕ

ϕ	factor₁	factor₂
1	0	0
2	$\pi \cos\theta_0$	$-\pi \sin\theta_0$
3	0	0
4	0	0

Table 4 Fourier – Harmonic functions factors

Then [i] is

$$\sum_{n=1}^{\infty} \left[B_n \int_{-\pi}^{\pi} \sin(kn\theta + \theta_0) \sin 2\theta \, d\theta + C_n \int_{-\pi}^{\pi} \cos(kn\theta + \theta_0) \sin 2\theta \, d\theta \right] =$$

$$= 0 + \underbrace{B_n \pi \cos\theta_0 - C_n \pi \sin\theta_0}_{\phi=2 \Rightarrow n = \frac{2}{k}} + 0 + 0 + 0 + \ldots = 0 \quad \forall \theta_0 \in [-\pi, \pi]$$

Now we can conclude that only two cases can fulfill the conditions:

1. If $k \geq 3$, then [i] is always fulfilled. The sections fulfilling this conditions are the sections with 3 or bigger order rotational symmetry.
2. If $k < 3$, then the following condition is required to get mechanical symmetry in the section:

$$B_n \cos\theta_0 = C_n \sin\theta_0 \quad \forall \theta_0 \in [-\pi, \pi]$$

A detailed examination of this condition:

$$\left. \begin{array}{l} B_n \cos\theta_0 = C_n \sin\theta_0 \quad \forall \theta_0 \in [-\pi, \pi] \\ A_n \cos\theta_n = B_n \\ A_n \sin\theta_n = C_n \end{array} \right\} \rightarrow A_n \cos\theta_n \cos\theta_0 = A_n \sin\theta_n \sin\theta_0$$

$$\left. \begin{array}{l} \tan\theta_0 = \tan\theta_n \quad \forall \theta_0 \in [-\pi, \pi] \\ \theta_n = \text{constant} \end{array} \right\} \Rightarrow \text{Impossible}$$

proves that there is only one way of fulfilling the constant MoI condition. The section has a rotational symmetry or order 3 or greater.

3.7 Theorem

Given a section delimited by a continuous, closed line (we will call it simple section) with mechanical symmetry (with constant moment of inertia respect of the rotation of the section around the center of mass of the section), we can say that the section has 3 or greater order rotational symmetry.

In other words:

*It is necessary and sufficient condition for a section to have **Mechanical Symmetry** having 3 or greater order rotational symmetry*

We show now an example to illustrate it. We use a periodic function with rotational symmetry: 1-, 2-, 3-, and 4-order rotational symmetry (say 1-order is no rotational symmetry) defined by:

$$\rho(\theta,\theta_0) = \cos\left(\frac{(\theta_0 + \theta - 1)k}{2}\right)^2 + \cos\left((\theta_0 + \theta - 2)k\right)^2$$

In the following table we show for each value of k:

- Section graph
- Graph for the functions: $\sin 2x$, $\rho(\theta,\theta_0) \equiv rol(x)$, $\rho^4(\theta,\theta_0)\sin 2x$
- Formula defining $\rho(\theta,\theta_0)$
- Values for θ_0 used to calculate
- Calculated value $\int_{-\pi}^{\pi} \rho^4(\theta,\theta_0)\sin 2x\, dx$

$k=1$

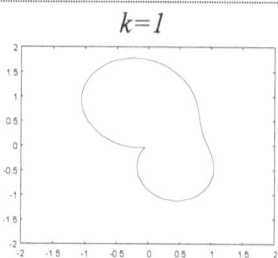

Mechanical Symmetry

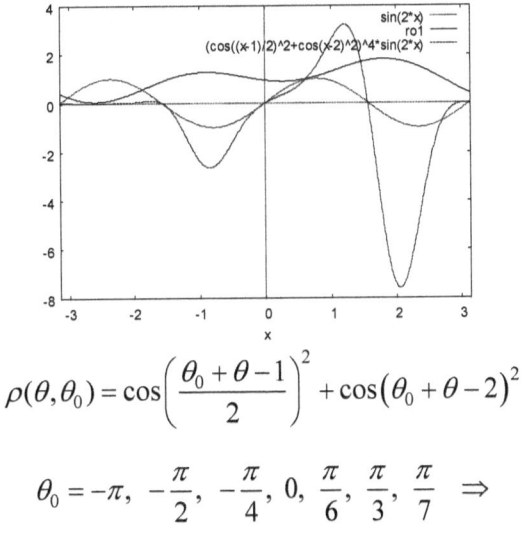

$$\rho(\theta,\theta_0) = \cos\left(\frac{\theta_0 + \theta - 1}{2}\right)^2 + \cos(\theta_0 + \theta - 2)^2$$

$$\theta_0 = -\pi,\ -\frac{\pi}{2},\ -\frac{\pi}{4},\ 0,\ \frac{\pi}{6},\ \frac{\pi}{3},\ \frac{\pi}{7} \Rightarrow$$

$$\int_{-\pi}^{\pi} (\rho(\theta,\theta_0))^4 \sin 2x\, dx = -5.0,\ 5.0,\ -7.4,\ -5.0,\ 3.9,\ 8.9,\ 2.7$$

$K=2$

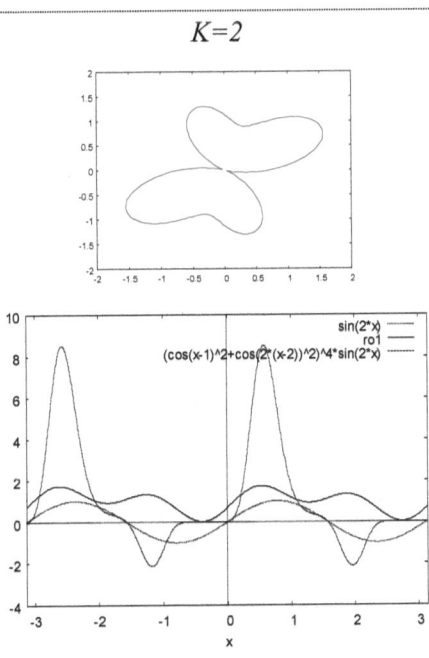

Mechanical Symmetry

$$\rho(\theta,\theta_0) = \cos\left((\theta_0+\theta-1)\right)^2 + \cos\left(2(\theta_0+\theta-2)\right)^2$$

$$\theta_0 = -\pi,\ -\frac{\pi}{2},\ -\frac{\pi}{4},\ 0,\ \frac{\pi}{6},\ \frac{\pi}{3},\ \frac{\pi}{7}\ \Rightarrow$$

$$\int_{-\pi}^{\pi} \left(\rho(\theta,\theta_0)\right)^4 \sin 2x\, dx = 7.4,\ -7.4,\ 1.0,\ 7.4,\ 2.8,\ -4.5,\ 3.8$$

K=3

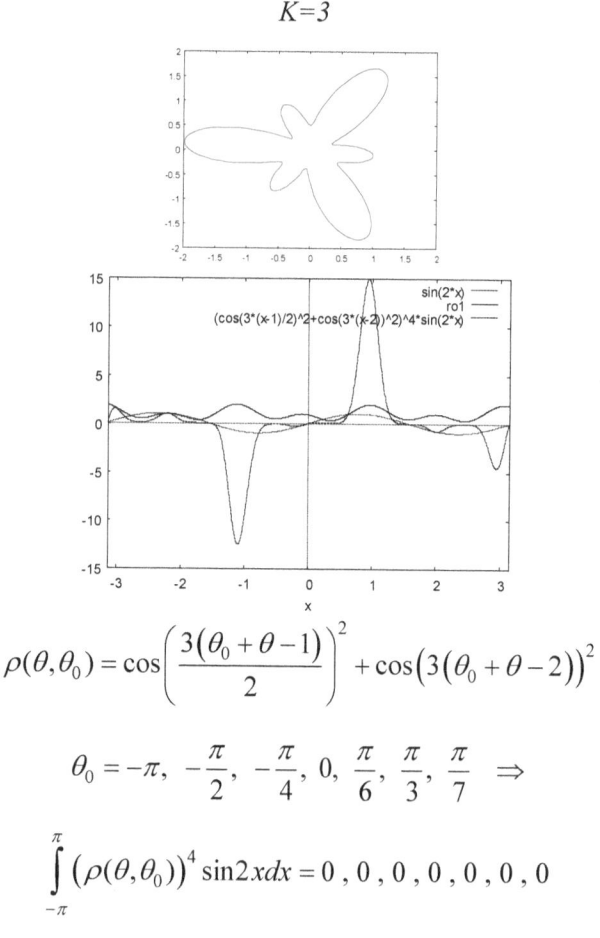

$$\rho(\theta,\theta_0) = \cos\left(\frac{3(\theta_0+\theta-1)}{2}\right)^2 + \cos\left(3(\theta_0+\theta-2)\right)^2$$

$$\theta_0 = -\pi,\ -\frac{\pi}{2},\ -\frac{\pi}{4},\ 0,\ \frac{\pi}{6},\ \frac{\pi}{3},\ \frac{\pi}{7}\ \Rightarrow$$

$$\int_{-\pi}^{\pi} \left(\rho(\theta,\theta_0)\right)^4 \sin 2x\, dx = 0,\ 0,\ 0,\ 0,\ 0,\ 0,\ 0$$

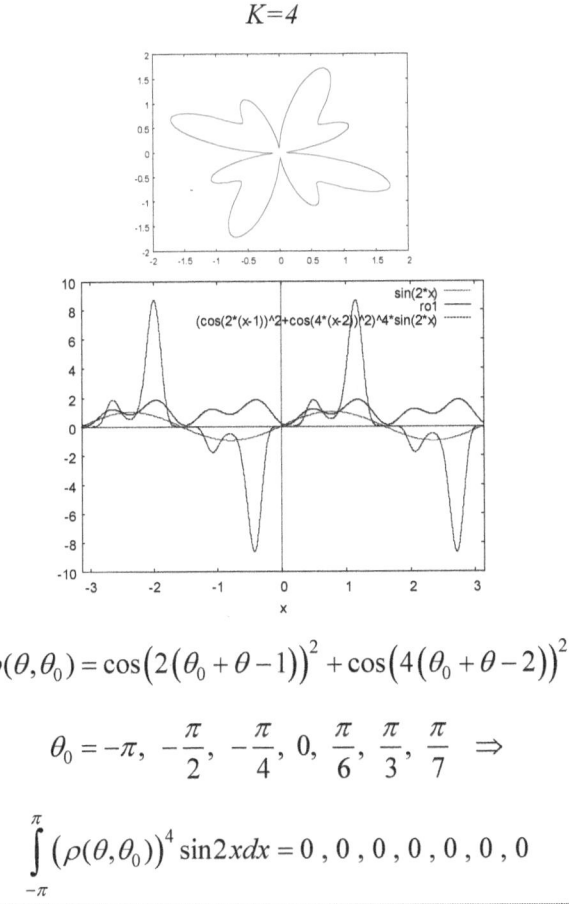

$$\rho(\theta,\theta_0) = \cos\left(2(\theta_0+\theta-1)\right)^2 + \cos\left(4(\theta_0+\theta-2)\right)^2$$

$$\theta_0 = -\pi,\ -\frac{\pi}{2},\ -\frac{\pi}{4},\ 0,\ \frac{\pi}{6},\ \frac{\pi}{3},\ \frac{\pi}{7}\ \Rightarrow$$

$$\int_{-\pi}^{\pi} (\rho(\theta,\theta_0))^4 \sin 2x\, dx = 0,0,0,0,0,0,0$$

Table 5 Fourier – Rotational Symmetry Samples

3.8 Why Should We Do It?

If a section has mechanical symmetry, then the neutral axis will be always a principal direction. This means that the deformation value will be independent from the direction of load application, or in a rotating part the position will not affect the value.

If the key factor when calculating a section is deformation or deflection (which is frequent), then using sections with mechanical symmetry simplifies the process and minimizes uncertainty.

When calculating vibrating parts or vibrations on elements (earthquakes), having MS sections simplifies the work.

No Answers

4

Mechanical Symmetry

4 No Answers

As far as my limited knowledge and my talent is not allowing to me to arrive further on this subjects, I will just write here what I know for the case; it can be useful for others, and I invite them to extend this to 3- or n-dimensional spaces.

4.1 Incoherence

From the equations [1] and [3], focusing on

$$\sum_{n=1}^{k} \cos 2\alpha_n \quad ; \quad \alpha_n = \frac{2\pi}{k} n + \alpha_0$$

we see what we call incoherence. For $k=1$ and for $k=2$ we have a variable result for the sum; for $k \geq 3$ it is constant.

$$k=1 \Rightarrow \sum_{n=1}^{1} \cos 2\alpha_n = \sum_{n=1}^{k} \cos\left(\frac{4\pi}{k} n + 2\alpha_0\right) = \cos(4\pi + 2\alpha_0) = \cos 2\alpha_0$$

$$k=2 \Rightarrow \sum_{n=1}^{2} \cos 2\alpha_n = \sum_{n=1}^{k} \cos\left(\frac{4\pi}{k} n + 2\alpha_0\right) =$$

$$= \cos(2\pi + 2\alpha_0) + \cos 2\alpha_0 = 2\cos 2\alpha_0$$

$$k \geq 3 \Rightarrow \sum_{n=1}^{k} \cos 2\alpha_n = 0$$

Obviously:

$$\sum_{n=1}^{k} \sin^2 \alpha_n = \sum_{n=1}^{k} \frac{(1-\cos 2\alpha_n)}{2} = \frac{1}{2}\left(\sum_{n=1}^{k} 1 - \sum_{n=1}^{k} \cos 2\alpha_n\right) = \frac{k}{2} - \sum_{n=1}^{k} \cos 2\alpha_n$$

Generalizing:

$$k=1 \Rightarrow \sum_{n=1}^{1} \sin^2 \alpha_n = \frac{1}{2} - \cos 2\alpha_0$$

$$k=2 \Rightarrow \sum_{n=1}^{2} \sin^2 \alpha_n = 1 - 2\cos 2\alpha_0$$

$$k \geq 3 \Rightarrow \sum_{n=1}^{k} \sin^2 \alpha_n = \frac{k}{2}$$

Mechanical Symmetry

$$\sum_{n=1}^{k}\sin^2\alpha_n \begin{cases} k<3 \Rightarrow \dfrac{k}{2} - k\cos 2\alpha_0 \\ k\geq 3 \Rightarrow \dfrac{k}{2} \end{cases} \forall\, k \in \mathbb{N} \quad ; \quad \alpha_n = \dfrac{2\pi}{k}n + \alpha_0$$

And for the cosine:

$$\sum_{n=1}^{k}\cos^2\alpha = \sum_{n=1}^{k}\left(1 - \sin^2\alpha\right) = k - \sum_{n=1}^{k}\sin^2\alpha$$

We arrive to

$$k=1 \Rightarrow \sum_{n=1}^{1}\cos^2\alpha_n = \dfrac{1}{2} + \cos 2\alpha_0$$

$$k=2 \Rightarrow \sum_{n=1}^{2}\cos^2\alpha_n = 1 + 2\cos 2\alpha_0$$

$$k\geq 3 \Rightarrow \sum_{n=1}^{k}\cos^2\alpha_n = \dfrac{k}{2}$$

$$\sum_{n=1}^{k}\cos^2\alpha_n \begin{cases} k<3 \Rightarrow \dfrac{k}{2} + k\cos 2\alpha_0 \\ k\geq 3 \Rightarrow \dfrac{k}{2} \end{cases} \forall\, k \in \mathbb{N} \quad ; \quad \alpha_n = \dfrac{2\pi}{k}n + \alpha_0$$

And obviously for every *k*:

$$\sum_{n=1}^{k}\cos^2\alpha_n + \sum_{n=1}^{k}\mathrm{sen}^2\alpha_n = k$$

4.2 Incoherent Sums

From what we have seen up to this moment we calculate the sum:

$$\sum_{n=1}^{k}\cos m\cdot\alpha_n$$

$$\alpha_n = \dfrac{2\pi}{k}n + \alpha_0 \qquad \forall\, k \in \mathbb{N}$$

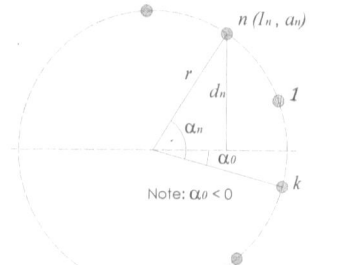

Fig. 27 Incoherent Sums

Using Euler's notation

Mechanical Symmetry

$$e^{\alpha i} = \cos\alpha + i\sin\alpha$$

We have:

$$\sum_{n=1}^{k}\cos m\alpha_n + i\sum_{n=1}^{k}\sin m\alpha_n$$

$$\sum_{n=1}^{k} e^{im\,\alpha_n} = \sum_{n=1}^{k} e^{im\left(\frac{2\pi}{k}n+\alpha_0\right)} = \sum_{n=1}^{k} e^{im\frac{2\pi}{k}n+m\alpha_0} = \sum_{n=1}^{k} e^{im\frac{2\pi}{k}n} e^{im\alpha_0}$$

This is the sum of a geometric progression from 1 to k

$$p = e^{im\frac{2\pi}{k}} \quad;\quad a_1 = e^{im\alpha_0}e^{im\frac{2\pi}{k}} \quad;\quad a_k = e^{im\alpha_0}e^{im\frac{2\pi}{k}k} = e^{im\alpha_0}e^{im\,2\pi}$$

$$S_k = \frac{a_1 - p\cdot a_k}{1-p}$$

Replacing

$$S_k = \frac{e^{im\alpha_0}e^{im\frac{2\pi}{k}} - e^{im\alpha_0}e^{im\frac{2\pi}{k}}\cdot e^{im\,2\pi}}{1-e^{im\frac{2\pi}{k}}} = \frac{\left(1-e^{im\,2\pi}\right)\cdot e^{im\alpha_0}e^{im\frac{2\pi}{k}}}{1-e^{im\frac{2\pi}{k}}} = \frac{0}{1-e^{im\frac{2\pi}{k}}}$$

Evaluating the denominator:

$$S_k = \frac{0}{1-e^{im\frac{2\pi}{k}}}$$

For the values of k:

$$k = 1$$

$$e^{i\frac{2\pi}{k}} = e^{i\,2\pi} = \cos 2\pi + i\sin 2\pi = 1 \Rightarrow 1-e^{i\frac{2\pi}{k}} = 1-1 = 0 \Rightarrow S_k \neq 0$$

$$k = m$$

$$e^{im\frac{2\pi}{k}} = e^{i\,2\pi} = \cos 2\pi + i\sin 2\pi = 1 \Rightarrow 1-e^{im\frac{2\pi}{k}} = 1-1 = 0 \Rightarrow S_k \neq 0$$

$$k > m$$

$$e^{im\frac{2\pi}{k}} = e^{i\,A\pi}; 0 \leq A \leq 1 \Rightarrow \cos A\pi + i\sin A\pi \neq 0 \Rightarrow 1-e^{im\frac{2\pi}{k}} \neq 0 \Rightarrow S_k = 0$$

Being clear that:
$$S_k = 0 \quad \forall\, k \in \mathbb{N}, k > m$$

Giving a deeper analysis to the values where $k \leq m$:
$$k = 1$$
$$\sum_{n=1}^{1} \cos m\alpha_n = \sum_{n=1}^{1} \cos m(2n\pi + \alpha_0) = \cos(2m\pi + m\alpha_0) = \cos m\alpha_0$$

$$k = 2 \Rightarrow$$
$$\sum_{n=1}^{2} \cos m\alpha_n = \sum_{n=1}^{2} \cos m(n\pi + \alpha_0) = \cos(m\pi + m\alpha_0) + \cos(m2\pi + m\alpha_0) =$$
$$= 2\cos m\alpha_0$$

$$k = m \Rightarrow \sum_{n=1}^{m} \cos m\alpha_n = \sum_{n=1}^{m} \cos m\left(\frac{2n\pi}{m} + \alpha_0\right) =$$
$$= \cos(2\pi + m\alpha_0) + \cos(4\pi + m\alpha_0) + \ldots + \cos(2m\pi + m\alpha_0) = m\cos m\alpha_0$$

$$k > m \Rightarrow \sum_{n=1}^{m} \cos m\alpha_n = 0$$

Concluding:

[k]
$$\sum_{n=1}^{k} \cos m \cdot \alpha_n \begin{cases} k \leq m \Rightarrow k\cos m\alpha_0 \\ k > m \Rightarrow 0 \end{cases} \forall\, k \in \mathbb{N}$$

$$\text{Being } \alpha_n = \frac{2\pi}{k} n + \alpha_0$$

Using It 5

Mechanical Symmetry

5. Using It

5.1 Regular Polygons

Regular polygons are probably the most commonly used figures with rotational symmetry. We will now calculate the MoI for regular polygons using mechanical symmetry formulas.

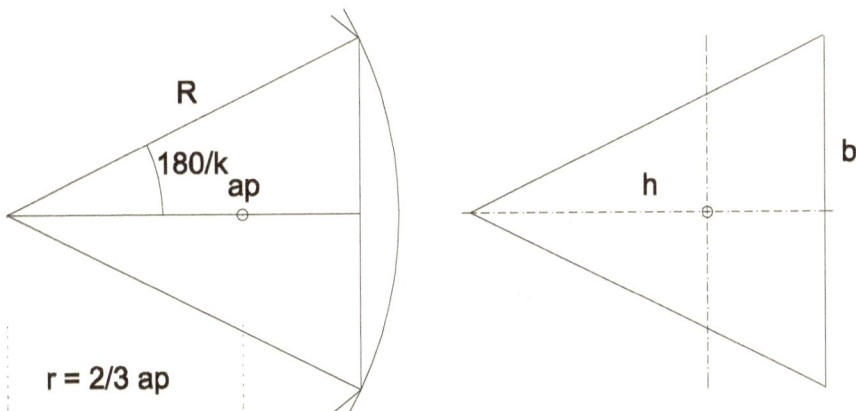

Fig. 28 Regular Polygons

Being:

R Circumcircle radius

$Ap=h$ Polygon's apothem

r Radius to center of mass of triangle with base L and height ap

$L=b$ Length of polygon side

The exact formula for MoI is:

[d] $$I_k = \frac{k}{2}\left(I_u + I_v + ar^2\right) \quad \forall k \in \mathbb{N} : k \geq 3$$

(In this case, we could ignore the condition $k \geq 3$)

Values for the summands in the formula are:

$$I_u = \frac{bh^3}{36} \; ; \; I_v = \frac{b^3 h}{48} \; ; \; h = ap = R\cos\frac{\pi}{k} \; ; \; b = 2ap\tan\frac{\pi}{k} = 2R\sin\frac{\pi}{k}$$

Mechanical Symmetry

First we calculate $I_u + I_v$ from R and k:

$$I_u + I_v = \frac{bh^3}{36} + \frac{b^3h}{48} = \frac{2R\sin\frac{\pi}{k} R^3 \left(\cos\frac{\pi}{k}\right)^3}{36} + \frac{8R^3\left(\sin\frac{\pi}{k}\right)^3 R\cos\frac{\pi}{k}}{48} =$$

$$= \frac{R^4}{18}\left(\sin\frac{\pi}{k}\left(\cos\frac{\pi}{k}\right)^3 + 3\left(\sin\frac{\pi}{k}\right)^3 \cos\frac{\pi}{k}\right) =$$

[l] $\quad\quad \dfrac{R^4 \sin\frac{2\pi}{k}}{36}\left(\left(\cos\frac{\pi}{k}\right)^2 + 3\left(\sin\frac{\pi}{k}\right)^2\right) = \dfrac{R^4 \sin\frac{2\pi}{k}}{36}\left(1 + 2\left(\sin\frac{\pi}{k}\right)^2\right) =$

[m] $\quad\quad = \dfrac{R^4 \sin\frac{2\pi}{k}}{36}\left(1 + 1 - \cos\frac{2\pi}{k}\right) = \dfrac{R^4 \sin\frac{2\pi}{k}}{36}\left(2 - \cos\frac{2\pi}{k}\right)$

Now the area:

[n] $\quad\quad a = \dfrac{bh}{2} = \dfrac{\left(R\cos\frac{\pi}{k}\right)\left(2R\sen\frac{\pi}{k}\right)}{2} = R^2 \cos\frac{\pi}{k}\sen\frac{\pi}{k} = \dfrac{R^2 \sin\frac{2\pi}{k}}{2}$

And distance to the center of mass:

$$r = \frac{2}{3}ap = \frac{2}{3}R\cos\frac{\pi}{k}$$

[o] $\quad\quad r^2 = \dfrac{4}{9}ap^2 = \dfrac{4}{9}R^2\left(\cos\frac{\pi}{k}\right)^2 = \dfrac{2}{9}R^2\left(1 - \cos\frac{2\pi}{k}\right)^2$

Substituting [l],[m],[n] and [o]:

$$I_k = \frac{k}{2}\left[\frac{R^4 \sin\frac{2\pi}{k}}{36}\left(2 - \cos\frac{2\pi}{k}\right) + \frac{R^2 \sin\frac{2\pi}{k}}{2}\frac{2}{9}R^2\left(1 - \cos\frac{2\pi}{k}\right)^2\right] =$$

$$= \frac{k}{2}\left[\frac{R^4 \sin\frac{2\pi}{k}}{36}\left(1 + 2sen^2\frac{\pi}{k}\right) + \frac{2R^4 \sin\frac{2\pi}{k}}{9}\cos^2\frac{\pi}{k}\right]$$

$$I_k = \frac{kR^4 \sin\frac{2\pi}{k}}{2}\left(\frac{1 + 2\sin^2\frac{\pi}{k}}{36} + \frac{8\cos^2\frac{\pi}{k}}{36}\right) =$$

$$= \frac{kR^4 \sin\frac{2\pi}{k}}{72}\left(1 + 2sen^2\frac{\pi}{k} + 8\cos^2\frac{\pi}{k}\right)$$

[p] $\quad\quad I_k = \dfrac{kR^4 \sin\frac{2\pi}{k}}{24}\left(1 + 2\cos^2\frac{\pi}{k}\right) = \dfrac{kR^4 \sin\frac{2\pi}{k}}{24}\left(2 + \cos\frac{2\pi}{k}\right)$

We calculate the value that maximizes the MoI with respect to the area (weight):

$$MaxMoI(k) = \frac{I_k}{ka} = \frac{R^2}{12}\left(2 + \cos\frac{2\pi}{k}\right)$$

Using this formula as a real variable continuous function over k and solving its derivative, we see that there are no real roots.

$$\frac{dMaxMoI(k)}{dk} = \frac{d}{dk}\left(\frac{R^2}{12}\left(2 + \cos\frac{2\pi}{k}\right)\right) = \pi\frac{R^2}{6k^2}\sin\frac{2\pi}{k} = 0$$

Therefore is in the infinite limit where the maximum is a circle.

Now we calculate the value for the MoI with the simplified formula.

$$I_{kaprox} = \frac{k}{2}\left(\frac{R^2 k \sin\frac{2\pi}{k}}{2}\left(\frac{2}{3}R\cos\frac{\pi}{k}\right)^2\right) = \frac{R^4 k \sin\frac{2\pi}{k} \cos^2\frac{\pi}{k}}{9}$$

With an accuracy

$$\Delta I_k = \frac{I_u + I_v}{ar^2} = \frac{\frac{R^4 \sin\frac{2\pi}{k}}{36}\left(2 - \cos\frac{2\pi}{k}\right)}{\frac{R^4 \sin\frac{2\pi}{k}}{36} 4\left(1 - \cos\frac{2\pi}{k}\right)^2} = \frac{2 - \cos\frac{2\pi}{k}}{4\left(1 - \cos\frac{\pi}{k}\right)^2} =$$

$$\Delta I_k = \frac{2 - \cos\frac{2\pi}{k}}{4\cos\frac{2\pi}{k} + 4} \qquad [q]$$

In any case, these formulas are not friendly at all. Let's see if we can get simpler formulas using some different parameters to define them.

Let's focus first on $I_u + I_v$

$$I_u = \frac{bh^3}{36} \; ; \; I_v = \frac{b^3 h}{48} \quad \Rightarrow \quad Iu + Iv = \frac{bh^3}{36} + \frac{b^3 h}{48}$$

Calculating now the area and distance to the center of mass:

$$a = \frac{bh}{2} \; ; \; r = \frac{2}{3}h$$

Replacing, we get

[r] $$I_k = \frac{k}{2}\left(\frac{bh^3}{4} + \frac{b^3h}{48} + \frac{bh}{2}\left(\frac{2}{3}h\right)^2\right) = \frac{k}{96}\left(12bh^3 + b^3h\right)$$

Now we calculate the MoI with the simplified formula and its accuracy.

$$I_k = \frac{k}{2}\left(\frac{bh}{2}\left(\frac{2}{3}h\right)^2\right) = \frac{kbh^3}{9}$$

$$\Delta I_k = \frac{\frac{(12bh^3 + b^3h)k}{96} - \frac{kbh^3}{9}}{\frac{kbh^3}{9}} = \frac{\frac{(12h^3 + b^2h)}{96} - \frac{h^3}{9}}{\frac{h^3}{9}} = \frac{4h^2 + 3b^2}{32h^2} = \frac{1}{8} + \frac{3b^2}{32h^2}$$

Showing that the simplified formula has a 12.5% error at least. In other words, it is not adequate. It is not a figure where the distance to the center of symmetry and area is more important than the MoI.

Substituting in [r] the values for b and h as a function of the radius R, polygon's apothem ap, side length L, and number of sides k, we get:

$$h = R\cos\tfrac{\pi}{k} \ ; \ b = 2R\sin\tfrac{\pi}{k}$$

[p] $$I_k = \frac{kR^4 \sin\tfrac{2\pi}{k}}{24}\left(3 - 2\sin^2\tfrac{\pi}{k}\right) = \frac{kR^4 \sin\tfrac{2\pi}{k}}{24}\left(2 + \cos\tfrac{2\pi}{k}\right)$$

Eq. 27 Regular Polygon MoI - From R and k

$$h = ap \ ; \ b = 2h \cdot \tan\tfrac{\pi}{k}$$

[s] $$I_k = \frac{k \cdot ap^4 \tan\tfrac{\pi}{k}}{12}\left[3 + \tan^2\tfrac{\pi}{k}\right]$$

Eq. 28 Regular Polygon MoI - From ap=h and k

$$h = \frac{b}{2\tan\tfrac{\pi}{k}} \ ; \ b = L$$

[t] $$I_k = \frac{k \cdot L^4}{192}\left(\frac{1}{\tan\tfrac{\pi}{k}} + \frac{3}{\tan^3\tfrac{\pi}{k}}\right) = \frac{k \cdot L^4}{192}\left(\cot\tfrac{\pi}{k} + 3\cot^3\tfrac{\pi}{k}\right)$$

Eq. 29 Regular Polygon MoI - From L and k

We can also get the formulas for value of the MoI as a function of the polygon's area A:

Mechanical Symmetry

[u] $$I_k = \frac{k \cdot A}{4}\left(h^2 + \frac{b^2}{12}\right)$$

[v] $$I_k = \frac{k \cdot A \cdot R^2}{12}\left(2 + \cos\tfrac{2\pi}{k}\right)$$

[w] $$I_k = \frac{k \cdot A \cdot ap^2}{12}\left(3 + \tan^2\tfrac{\pi}{k}\right)$$

[x] $$I_k = \frac{k \cdot A \cdot L^2}{48}\left(1 + 3\cot^2\tfrac{\pi}{k}\right)$$

Eq. 30 Regular Polygon MoI – With Area

It can be verified that for big enough k values the formula is equal to the circle MoI. Expressed in mathematical terms:

$$\lim_{k\to\infty} I_k = \lim_{k\to\infty} \frac{kAR^2}{12}\left(2 + \cos\tfrac{2\pi}{k}\right) = \frac{R^4}{12}\lim_{k\to\infty}\left[k\cos\tfrac{\pi}{k}\sin\tfrac{\pi}{k}\left(2 + \cos\tfrac{2\pi}{k}\right)\right] = \frac{\pi R^4}{4}$$

Now we know a regular polygon's MoI, and we can formulate it as a function of the possible parameters. We also know that the optimal use of a structural polygonal section (maximizing MoI to area ratio) is a circular shape.

5.2 Circles and Computers

We are going to give a practical use to this knowledge. To do it, we will give an optimal answer to the problem of a circular section discretization to be used on mechanical calculations.

When discretizing or modeling, it is not always possible to get exact results for all the physical magnitudes involved.

In this case, we will use a regular polygon as model for a circle. We see in the circle 3 in *Fig. 29* that depending on the choice of polygon (making the circle incircle or circumcircle for the polygon), we have a different radius (r o R) and different side (c or C). If we choose the polygon inside the circle, the vertex of the polygon will have the same stress due to bending than the real circle, but the overall deflection of the element will not be the same, and vice versa.

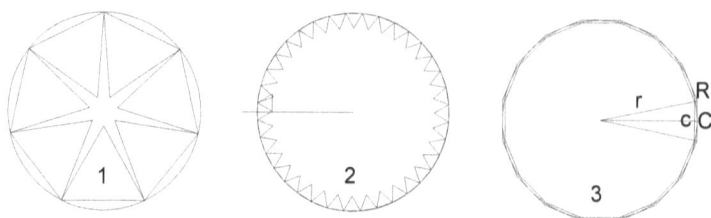

Fig. 29 Circles and Computers – Discretization

There are three magnitudes implied, and three section properties related to them:

1. Axial elongation → Area.

$$\sigma = \frac{N}{\Omega}; \delta = L\varepsilon; \sigma = E\varepsilon \Rightarrow \delta = L\frac{N}{E\Omega} \rightarrow \Omega$$

2. Bending deflection → MoI.

$$\delta = \delta(M, E, I) \rightarrow I$$

3. Maximum stress due to bending → W Modulus. Defined as (y being the longest distance to neutral axis).

$$\sigma = \frac{M}{I}y \; ; \; W = \frac{I}{y} \rightarrow W$$

Stress due to axial loads is not mentioned because it is included in condition 1. Shear stress and deformation are not included because they do not depend on only MoI.

We know the exact values for a circular section and exact formulas for a regular polygon section. With this data we can obtain the values for k, fulfilling some of the above mentioned conditions.

Fulfilling conditions 1 and 2:

Being R_c circle's radius.

$$kR^2 \cos\tfrac{\pi}{k} \sin\tfrac{\pi}{k} = \pi R_c^2$$

Eq. 31 Circles and Computers – Condition 1 Equal Area

$$\frac{kR^4 \sin\tfrac{2\pi}{k}}{24}\left(2 + \cos\tfrac{2\pi}{k}\right) = \frac{\pi R_c^4}{4}$$

Eq. 32 Circles and Computers – Condition 2 Equal MoI

Mechanical Symmetry

Solving the system:

$$R^2 = \frac{\pi R_c^2}{k\cos\frac{\pi}{k}\sin\frac{\pi}{k}} \rightarrow \frac{k\left(\frac{\pi R_c^2}{k\cos\frac{\pi}{k}\sin\frac{\pi}{k}}\right)^2 \sin\frac{2\pi}{k}}{24}(2+\cos\frac{\pi}{k}) = \frac{\pi R_c^4}{4}$$

$$\frac{4\pi R_c^4}{24k\sin\frac{2\pi}{k}}(2+\cos\frac{2\pi}{k}) = \frac{\pi R_c^4}{4}$$

$$\frac{2}{3k\sin\frac{2\pi}{k}}(2+\cos\frac{2\pi}{k}) = 1$$

We can see from this result that there is no solution for the system. In other words, the radius making the area of the polygon equal to that of the circle is not the same as the one making equal the MoI.

Fulfilling conditions 2 and 3:

$$\frac{k R^4 \sin\frac{2\pi}{k}}{24}(2+\cos\frac{2\pi}{k}) = \frac{\pi R_c^4}{4}$$

Eq. 32 Circles and Computers – Condition 2 Equal MoI

$$\frac{\frac{k R^4 \sin\frac{2\pi}{k}}{24}(2+\cos\frac{2\pi}{k})}{R} = \frac{\frac{\pi R_c^4}{4}}{R_c}$$

Eq. 33 Circles and Computers – Condition 3 Equal W Modulus

Solving the system:

$$R_c = R \cdot \sqrt[4]{\frac{k(2+\cos\frac{2\pi}{k})\sin\frac{2\pi}{k}}{6\pi}}$$

$$\frac{(\cos\frac{2\pi}{k}+2)\sin\frac{2\pi}{k} k R^3}{6} = R^3 \cdot \sqrt[4]{\frac{\pi k^3 (2+\cos\frac{2\pi}{k})^3 \sin^3\frac{2\pi}{k}}{216}}$$

We arrive again at a system with no practical solution.

From what we have seen, we can conclude that we cannot get values for R and k by giving exact solution for more than one condition.

As a simple solution, but not a bad one, we can use the average value between these three (knowing that they are close for big values of k), obtaining a similar error for the resulting values in axial strain, strain due to bending and bending stress.

$$R_1 = \sqrt{\frac{\pi R_c^{\,2}}{k\cos\frac{\pi}{k}\sin\frac{\pi}{k}}} = R_c\sqrt{\frac{\pi}{k\cos\frac{\pi}{k}\sin\frac{\pi}{k}}}$$

$$R_2 = \sqrt[4]{\frac{\pi R_c^{\,4}}{\frac{k\sin\frac{2\pi}{k}}{24}\left(2+\cos\frac{2\pi}{k}\right)}} = R_c\sqrt[4]{\frac{6\pi}{k\sin\frac{2\pi}{k}\left(2+\cos\frac{2\pi}{k}\right)}}$$

$$R_3 = \sqrt[3]{\frac{\pi R_c^{\,3}}{\frac{k\sin\frac{2\pi}{k}}{24}\left(2+\cos\frac{2\pi}{k}\right)}} = R_c\sqrt[3]{\frac{6\pi}{k\sin\frac{2\pi}{k}\left(2+\cos\frac{2\pi}{k}\right)}}$$

$$R_p = \frac{R_1 + R_2 + R_3}{3}$$

Being:
R_i each condition's radius
R_p proposed radius

We show now the complete development of this proposal with a workbook for the Maxima software.

(%i1) assume(k≥ 3);

(%o1)[$k > 2$]

---

1. Conditions 1 and 2

Condition 1 – equal area

(%i2) aux:(%pi*Rc^2)/(k*cos(%pi/k)*sin(%pi/k));

$$(\%o2) \frac{\pi Rc^2}{\cos\left(\frac{\pi}{k}\right)\sin\left(\frac{\pi}{k}\right)k}$$

<u>Condition 2 – equal MoI</u>

```
(%i3) poli:k*aux^2*sin(2*%pi/k)/24*(2+cos(2*%pi/k));
```

$$(\%o3) \frac{\pi^2\left(\cos\left(\frac{2\pi}{k}\right)+2\right)\sin\left(\frac{2\pi}{k}\right)Rc^4}{24\cos\left(\frac{\pi}{k}\right)^2\sin\left(\frac{\pi}{k}\right)^2 k}$$

```
(%i4) circul:%pi*Rc^4/4;
```

$$(\%o4) \frac{\pi Rc^4}{4}$$

```
(%i5) poli=circul;
```

$$(\%o5) \frac{\pi^2\left(\cos\left(\frac{2\pi}{k}\right)+2\right)\sin\left(\frac{2\pi}{k}\right)Rc^4}{24\cos\left(\frac{\pi}{k}\right)^2\sin\left(\frac{\pi}{k}\right)^2 k} = \frac{\pi Rc^4}{4}$$

```
(%i6) res:solve([%], [k]);
```

$$(\%o6) [k = \frac{\left(\pi\cos\left(\frac{2\pi}{k}\right)+2\pi\right)\sin\left(\frac{2\pi}{k}\right)}{6\cos\left(\frac{\pi}{k}\right)^2\sin\left(\frac{\pi}{k}\right)^2}]$$

<u>0 < (k/k) < 1</u>

```
(%i7) k/(rhs(res[1]));
```

$$(\%o7) \frac{6\cos\left(\frac{\pi}{k}\right)^2\sin\left(\frac{\pi}{k}\right)^2 k}{\left(\pi\cos\left(\frac{2\pi}{k}\right)+2\pi\right)\sin\left(\frac{2\pi}{k}\right)}$$

```
(%i8) eq:trigrat(%);
```

$$(\%o8) \frac{3\sin\left(\frac{2\pi}{k}\right)k}{2\pi\cos\left(\frac{2\pi}{k}\right)+4\pi}$$

```
(%i9) wxplot2d([eq,1], [k,2.1,50])$
```

Mechanical Symmetry

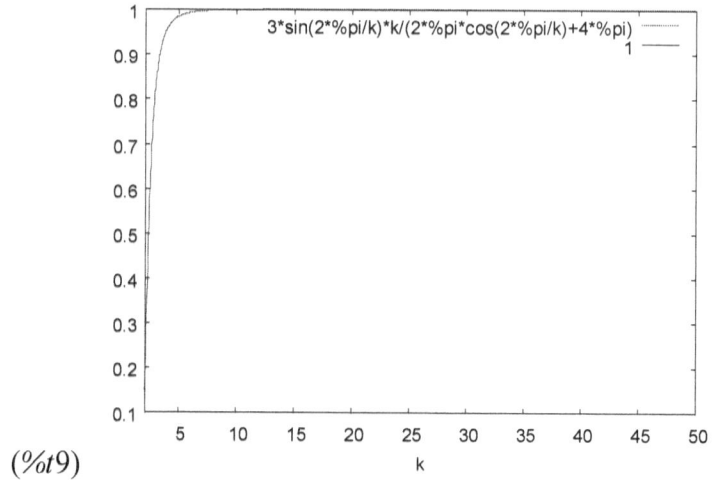

(%t9)

Fig. 30 Circles and Computers – Conditions 1 & 2

(%i10) wxplot2d([eq,1], [k,2.1,4])$

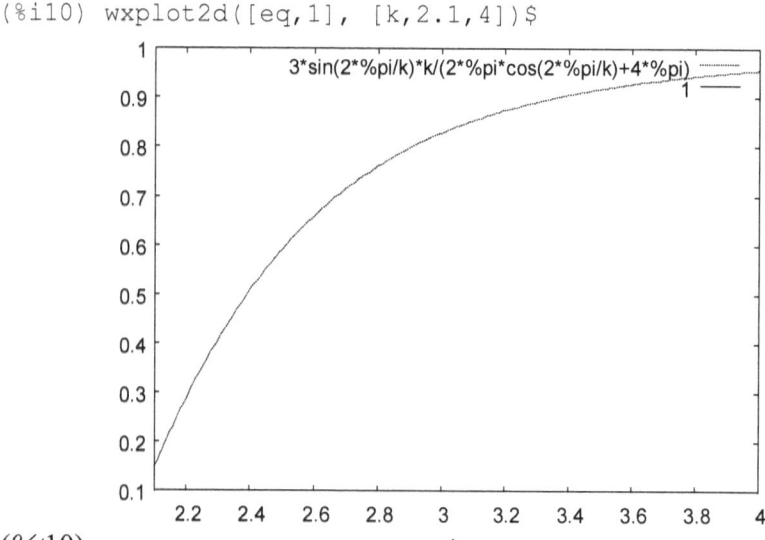

(%t10)

Fig. 31 Circles and Computers – Conditions 1 & 2 (detail)

---

2. Conditions 2 and 3

Condition 2 – equal MoI

(%i11) poli:k*R^4*sin(2*%pi/k)/24*(2+cos(2*%pi/k));

$$(\%o11) \frac{\left(\cos\left(\frac{2\pi}{k}\right)+2\right)\sin\left(\frac{2\pi}{k}\right)k\,R^4}{24}$$

(%i12) circul:%pi*Rc^4/4;

$$(\%o12) \frac{\pi\,Rc^4}{4}$$

(%i13) eq1:poli=circul;

$$(\%o13) \frac{\left(\cos\left(\frac{2\pi}{k}\right)+2\right)\sin\left(\frac{2\pi}{k}\right)k\,R^4}{24} = \frac{\pi\,Rc^4}{4}$$

(%i14) res1:solve([%], [Rc]);

$$(\%o14) \begin{bmatrix} Rc = \dfrac{i\left(\cos\left(\frac{2\pi}{k}\right)+2\right)^{\frac{1}{4}}\sin\left(\frac{2\pi}{k}\right)^{\frac{1}{4}}k^{\frac{1}{4}}R^{\frac{1}{4}}}{6}\pi^{\frac{1}{4}}, \\[2ex] Rc = -\dfrac{\left(\cos(frac2\pi k)+2\right)^{\frac{1}{4}}\sin\left(\frac{2\pi}{k}\right)^{\frac{1}{4}}k^{\frac{1}{4}}R^{\frac{1}{4}}}{6}\pi^{\frac{1}{4}}, \\[2ex] Rc = -\dfrac{i\left(\cos\left(\frac{2\pi}{k}\right)+2\right)^{frac14}\sin\left(\frac{2\pi}{k}\right)^{\frac{1}{4}}k^{\frac{1}{4}}R^{\frac{1}{4}}}{6}\pi^{\frac{1}{4}}, \\[2ex] Rc = \dfrac{\left(\cos\left(\frac{2\pi}{k}\right)+2\right)^{\frac{1}{4}}\sin\left(\frac{2\pi}{k}\right)^{\frac{1}{4}}k^{\frac{1}{4}}R^{\frac{1}{4}}}{6}\pi^{\frac{1}{4}} \end{bmatrix}$$

(%i15) Rc:rhs(res1[4]);

$$(\%o15) \frac{\left(\cos\left(\frac{2\pi}{k}\right)+2\right)^{\frac{1}{4}}\sin\left(\frac{2\pi}{k}\right)^{\frac{1}{4}}k^{\frac{1}{4}}R^{\frac{1}{4}}}{6}\pi^{\frac{1}{4}}$$

Condition 3 – equal maximum stress

(%i16) poli/R=circul/Rc;

$$(\%o16) \frac{\left(\cos\left(\frac{2\pi}{k}\right)+2\right)\sin\left(\frac{2\pi}{k}\right)k\,R^3}{24} = \frac{6^{\frac{1}{4}}\pi^{\frac{5}{4}}Rc^4}{4\left(\cos\left(\frac{2\pi}{k}\right)+2\right)^{\frac{1}{4}}\sin\left(\frac{2\pi}{k}\right)^{\frac{1}{4}}k^{\frac{1}{4}}R}$$

(%i17) eq2:ev(%, nouns);

$$(\%o17) \frac{\left(\cos\left(\frac{2\pi}{k}\right)+2\right)\sin\left(\frac{2\pi}{k}\right)kR^3}{24} = \frac{\pi^{-\frac{1}{4}}\left(\cos\left(\frac{2\pi}{k}\right)+2\right)^{\frac{3}{4}}\sin\left(\frac{2\pi}{k}\right)^{\frac{3}{4}}k^{\frac{3}{4}}R^3}{46^{\frac{3}{4}}}$$

(%i18) R:1;

(%o18) 1

(%i19) res2:solve([eq2], [k]);

$$(\%o19)[k = \frac{6\pi^{\frac{1}{4}}\left(\cos\left(\frac{2\pi}{k}\right)+2\right)^{\frac{3}{4}}k^{\frac{3}{4}}}{\left(6^{\frac{3}{4}}\cos\left(\frac{2\pi}{k}\right)+2\cdot6^{\frac{3}{4}}\right)\sin\left(rac2\pi k\right)^{\frac{1}{4}}}]$$

<u>0 < (k/k) < 1</u>
(%i20) eq:k/rhs(res2[1]);

$$(\%o20)\frac{\left(6^{\frac{3}{4}}\cos\left(\frac{2\pi}{k}\right)+2\cdot6^{\frac{3}{4}}\right)\sin\left(\frac{2\pi}{k}\right)^{\frac{1}{4}}k^{\frac{1}{4}}}{6\pi^{\frac{1}{4}}\left(\cos\left(\frac{2\pi}{k}\right)+2\right)^{\frac{3}{4}}}$$

(%i21) wxplot2d([eq,1], [k,2.1,25])$

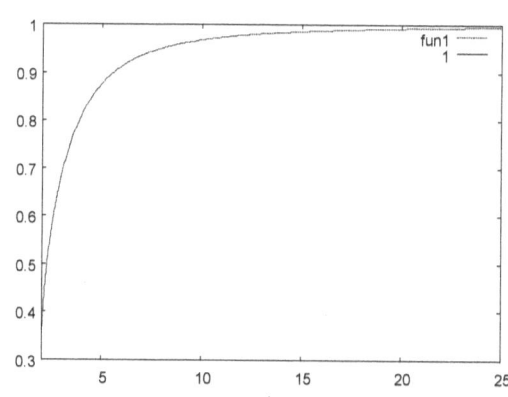

Fig. 32 Circles and Computers – Conditions 2 & 3

(%t21)

(%i22) wxplot2d([eq,1], [k,5000,50000])$

Mechanical Symmetry

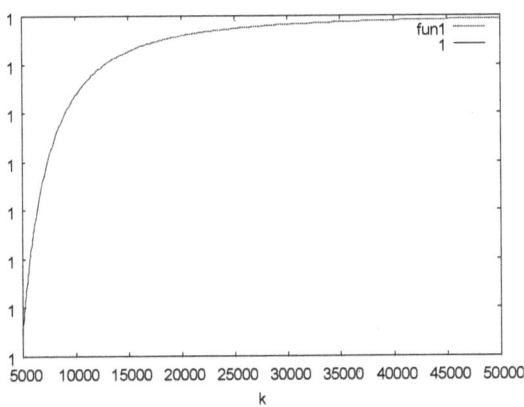

Fig. 33 Circles and Computers – Conditions 2 & 3 (detail)

(%t22)

---

3. Proposed Radius

(%i23) kill(all);

(%o0)*done*

Condition 1 – equal area
(%i1) eq1:k*R1^2*cos(%pi/k)*sin(%pi/k)=%pi*Rc^2;

$$(\%o1)\cos\left(\frac{\pi}{k}\right)\sin\left(\frac{\pi}{k}\right)k\,R1^2 = \pi\,Rc^2$$

(%i2) solve(%,R1);

$$(\%o2)\left[\begin{array}{l} R1 = -\sqrt{\pi}\sqrt{\dfrac{1}{\cos\left(\frac{\pi}{k}\right)\sin\left(\frac{\pi}{k}\right)k}}\,Rc, \\ R1 = \sqrt{\pi}\sqrt{\dfrac{1}{\cos\left(\frac{\pi}{k}\right)\sin\left(\frac{\pi}{k}\right)k}}\,Rc \end{array}\right]$$

(%i3) R1:rhs(%[2]);

$$(\%o3)\sqrt{\pi}\sqrt{\dfrac{1}{\cos\left(\frac{\pi}{k}\right)\sin\left(\frac{\pi}{k}\right)k}}\,Rc$$

Condition 2 — equal MoI
(%i4) poli:k*R^4*sin(2*%pi/k)/24*(2+cos(2*%pi/k));

(%o4) $\dfrac{\left(\cos\left(\frac{2\pi}{k}\right)+2\right)\sin\left(\frac{2\pi}{k}\right)k\,R^4}{24}$

(%i5) `circul:%pi*Rc^4/4;`

(%o5) $\dfrac{\pi\,Rc^4}{4}$

(%i6) `eq2:poli=circul;`

(%o6) $\dfrac{\left(\cos\left(\frac{2\pi}{k}\right)+2\right)\sin\left(\frac{2\pi}{k}\right)k\,R^4}{24}=\dfrac{\pi\,Rc^4}{4}$

(%i7) `solve([eq2], [R]);`

(%o7) $\left[R=\dfrac{6^{\frac{1}{4}}\pi^{\frac{1}{4}}i\left(\frac{1}{\sin\left(\frac{2\pi}{k}\right)k}\right)^{\frac{1}{4}}Rc}{\left(\cos\left(\frac{2\pi}{k}\right)+2\right)^{\frac{1}{4}}},\;R=-\dfrac{6^{\frac{1}{4}}\pi^{\frac{1}{4}}\left(\frac{1}{\sin\left(\frac{2\pi}{k}\right)k}\right)^{\frac{1}{4}}Rc}{\left(\cos\left(\frac{2\pi}{k}\right)+2\right)^{\frac{1}{4}}},\;R=-\dfrac{6^{\frac{1}{4}}\pi^{\frac{1}{4}}i\left(\frac{1}{\sin\left(\frac{2\pi}{k}\right)k}\right)^{\frac{1}{4}}Rc}{\left(\cos\left(\frac{2\pi}{k}\right)+2\right)^{\frac{1}{4}}},\;R=\dfrac{6^{\frac{1}{4}}\pi^{\frac{1}{4}}\left(\frac{1}{\sin\left(\frac{2\pi}{k}\right)k}\right)^{\frac{1}{4}}Rc}{\left(\cos\left(\frac{2\pi}{k}\right)+2\right)^{\frac{1}{4}}}\right]$

(%i8) `R2:rhs(%[4]);`

(%o8) $\dfrac{6^{\frac{1}{4}}\pi^{\frac{1}{4}}\left(\frac{1}{\sin\left(\frac{2\pi}{k}\right)k}\right)^{\frac{1}{4}}Rc}{\left(\cos\left(\frac{2\pi}{k}\right)+2\right)^{\frac{1}{4}}}$

Condition 3 – equal maximum stress

(%i9) `poli/R=circul/Rc;`

(%o9) $\dfrac{\left(\cos\left(\frac{2\pi}{k}\right)+2\right)\sin\left(\frac{2\pi}{k}\right)k\,R^3}{24}=\dfrac{\pi\,Rc^3}{4}$

(%i10) `eq3:ev(%, nouns);`

(%o10) $\dfrac{\left(\cos\left(\frac{2\pi}{k}\right)+2\right)\sin\left(\frac{2\pi}{k}\right)k\,R^3}{24} = \dfrac{\pi\,Rc^3}{4}$

(%i11) solve([eq3], [R]);

(%o11) $\left[R = \dfrac{\left(\sqrt{3}\,6^{\frac{1}{3}}\pi^{\frac{1}{3}}i - 6^{\frac{1}{3}}\pi^{\frac{1}{3}}\right)Rc}{2\left(\cos\left(\frac{2\pi}{k}\right)+2\right)^{\frac{1}{3}}\sin\left(\frac{2\pi}{k}\right)^{ac13}k^{\frac{1}{3}}}, \right.$

$R = -\dfrac{\left(\sqrt{3}\,6^{\frac{1}{3}}\pi^{\frac{1}{3}}i + 6^{\frac{1}{3}}\pi^{\frac{1}{3}}\right)Rc}{2\left(\cos\left(\frac{2\pi}{k}\right)+2\right)^{\frac{1}{3}}\sin\left(frac2\pi k\right)^{\frac{1}{3}}k^{\frac{1}{3}}},$

$\left. R = \dfrac{6^{\frac{1}{3}}\pi^{\frac{1}{3}}Rc}{\left(\cos\left(\frac{2\pi}{k}\right)+2\right)^{\frac{1}{3}}}\sin\left(\frac{2\pi}{k}\right)^{\frac{1}{3}}k^{\frac{1}{3}} \right]$

(%i12) R3:rhs(%[3]);

(%o12) $\dfrac{6^{\frac{1}{3}}\pi^{\frac{1}{3}}Rc}{\left(\cos\left(\frac{2\pi}{k}\right)+2\right)^{\frac{1}{3}}}\sin\left(\dfrac{2\pi}{k}\right)^{\frac{1}{3}}k^{\frac{1}{3}}$

To draw the circle's radius Rc and the radius of the three approximations, R1, R2, and R3

(%i13) Rc:1;

(%o13) 1

(%i14) [Rc,ev(R1,nouns),ev(R2, nouns),ev(R3, nouns)];

(%o14) $\left[1,\ \sqrt{\pi}\sqrt{\dfrac{1}{\cos\left(\frac{\pi}{k}\right)\sin\left(\frac{\pi}{k}\right)k}},\right.$

$\dfrac{6^{\frac{1}{4}}\pi^{\frac{1}{4}}\left(\dfrac{1}{\sin\left(\frac{2\pi}{k}\right)k}\right)^{\frac{1}{4}}}{\left(\cos\left(\frac{2\pi}{k}\right)+2\right)^{\frac{1}{4}}},$

$\left. \dfrac{6^{\frac{1}{3}}\pi^{\frac{1}{3}}}{\left(\cos\left(\frac{2\pi}{k}\right)+2\right)^{\frac{1}{3}}}\sin\left(\dfrac{2\pi}{k}\right)^{\frac{1}{3}}k^{\frac{1}{3}} \right]$

```
(%i15) plot2d(%, [k,3,20],[y,0,2*Rc],[legend,"Rc","R1
","R2","R3"])$
```

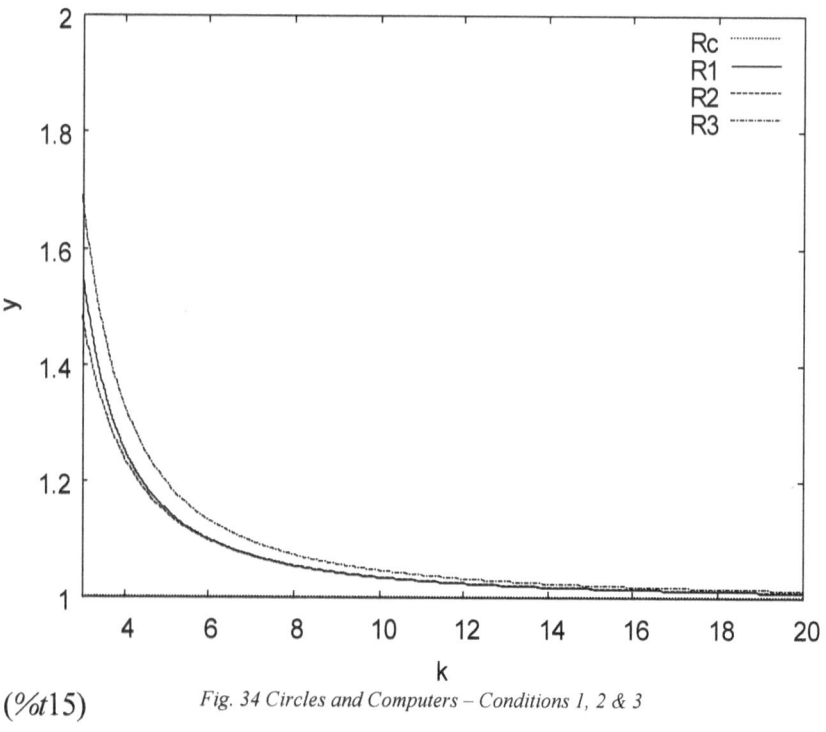

(%t15)   *Fig. 34 Circles and Computers – Conditions 1, 2 & 3*

We can see now why we will never be able to get solutions fulfilling two conditions simultaneously: three lines act as asymptotes for the rest. But we can get very close using the values from the following table.

With *R1* the radius fulfilling condition 1 (axial stress and elongation), *R2* fulfilling condition 2 (bending deflection), and *R3* fulfilling condition 3 (maximum stress due to bending), in the next table we find the values matching the properties with the one for the original circle. For example, a hexagon with a 1,0975 radius circumcircle has the same bending deflection as a circle with radius 1.

$$R_1 = F1 \cdot R_c$$
$$R_2 = F2 \cdot R_c$$
$$R_3 = F2 \cdot R_c$$

Mechanical Symmetry

Table 6 Circles and Computers – Individual Factors

k	F1	F2	F3
3	1,5551	1,4830	1,6912
4	1,2533	1,2389	1,3307
5	1,1495	1,1447	1,1974
6	1,0996	1,0975	1,1321
7	1,0715	1,0704	1,0950
8	1,0539	1,0533	1,0717
9	1,0422	1,0418	1,0561
10	1,0339	1,0337	1,0451
12	1,0233	1,0232	1,0311
14	1,0170	1,0170	1,0227
16	1,0130	1,0130	1,0173
18	1,0102	1,0102	1,0137
20	1,0083	1,0083	1,0110
30	1,0037	1,0037	1,0049
40	1,0021	1,0021	1,0027
50	1,0013	1,0013	1,0018
100	1,0003	1,0003	1,0004
360	1,0000	1,0000	1,0000

With the radius fulfilling simultaneously the conditions 1, 2, and 3; 1 and 2; 2 and 3; and 1 and 3, respectively:

$$R_{123} = F123 \cdot R_c$$
$$R_{12} = F12 \cdot R_c$$
$$R_{23} = F23 \cdot R_c$$
$$R_{13} = F13 \cdot R_c$$

In the next table are the values for these equivalencies. The error for the approximation is in the column *ErrXXX*, adjacent to the factor *FXXX*. If we replace a circle with radius 1 by a triangle with 1.5191 radius circumcircle, the difference between the exact and approximate valued for the area and the MoI will be less or equal than 2.37%. If we replace a circle with radius 1 by a dodecagon with 1,0259 radius circumcircle, the error for the three properties(area, MoI and W modulus) will be less than or equal than 0.51%.

Table 7 Circles and Computers – Combined Factors and Errors

k	F123	Err123	F12	Err12	F23	Err23	F13	Err13
3	1,5764	7,28%	1,5191	2,37%	1,5871	6,56%	1,6231	4,19%
4	1,2743	4,42%	1,2461	0,58%	1,2848	3,57%	1,2920	2,99%
5	1,1638	2,88%	1,1471	0,21%	1,1710	2,25%	1,1734	2,04%
6	1,1098	2,01%	1,0986	0,10%	1,1148	1,55%	1,1159	1,46%
7	1,0790	1,48%	1,0709	0,05%	1,0827	1,13%	1,0832	1,08%
8	1,0596	1,14%	1,0536	0,03%	1,0625	0,87%	1,0628	0,84%
9	1,0467	0,90%	1,0420	0,02%	1,0490	0,68%	1,0491	0,66%
10	1,0376	0,73%	1,0338	0,01%	1,0394	0,55%	1,0395	0,54%
12	1,0259	0,51%	1,0233	0,01%	1,0271	0,38%	1,0272	0,38%
14	1,0189	0,37%	1,0170	0,00%	1,0198	0,28%	1,0199	0,28%
16	1,0144	0,29%	1,0130	0,00%	1,0151	0,21%	1,0152	0,21%
18	1,0114	0,23%	1,0102	0,00%	1,0119	0,17%	1,0120	0,17%
20	1,0092	0,18%	1,0083	0,00%	1,0097	0,14%	1,0097	0,14%
30	1,0041	0,08%	1,0037	0,00%	1,0043	0,06%	1,0043	0,06%
40	1,0023	0,05%	1,0021	0,00%	1,0024	0,03%	1,0024	0,03%
50	1,0015	0,03%	1,0013	0,00%	1,0015	0,02%	1,0015	0,02%
100	1,0004	0,01%	1,0003	0,00%	1,0004	0,01%	1,0004	0,01%
360	1,0000	0,00%	1,0000	0,00%	1,0000	0,00%	1,0000	0,00%

Lines mark the values for the error lower than 1% and 1‰.

Formula Compilation

Mechanical Symmetry

6. Formula Compilation

6.1 Sections with Mechanical Symmetry

Formula for any section, regardless of its particular shape, with mechanical symmetry.

Mechanical Symmetry

Mechanical Symmetry

$$\forall k \in \mathbb{N} : k \geq 3$$

Being:
a Element area
I_u, I_v Element Principal Moments of Inertia

Exact Formula

Area	$k \cdot a$	
Center of Mass	(x_c, y_c) Symmetry or Rotation Center	
Moment of Inertia	$\begin{cases} I_k = I_x = I_y = \dfrac{k}{2}\left(I_u + I_v + a r^2\right) \\ I_{xy} = 0 \end{cases}$	[d]
Radius of Gyration	$i_k = i_x = i_y = \sqrt{\dfrac{k}{2}\left(\dfrac{I_u + I_v}{a} + r^2\right)}$	[f]
Polar Moment	$I_P = k\left(I_u + I_v + a r^2\right)$	[g]

Approximate Formula

Area	$k \cdot a$	
Center of Mass	(x_c, y_c) Symmetry or Rotation Center	
Moment of Inertia	$\begin{cases} I_{kap} = I_x = I_y = \dfrac{k a r^2}{2} \\ I_{xy} = 0 \end{cases}$	[a]
Radius of Gyration	$i_{kap} = i_{xap} = i_{yap} = r\sqrt{\dfrac{k}{2a}}$	[j]
Polar Moment	$I_{Pap} = k a r^2$	[k]
Accuracy	$\Delta I_k = \dfrac{I_u + I_v}{a \cdot r^2}$	[i]

Mechanical Symmetry

6.2 Regular Polygons

All formulas for an axis passing through the center of symmetry.

For all cases:

$$I_{xy} = 0 \Rightarrow I_u = I_v = I_x = I_y = I_k$$

$$i_{xy} = 0 \Rightarrow i_u = i_v = i_x = i_y = i_k$$

$$\Delta I_k = \frac{I_k - I_{kap}}{I_{kap}}$$

The parameters we have used to define a polygon are:

    k – Number of sides
    R – Circumcircle radius
    b – Side length
    h – Apothem or incircle radius

$\alpha = \frac{2\pi}{k}$ (rad)

Area: $\Omega = \dfrac{kbh}{2}$

Perimeter: $P = kb$

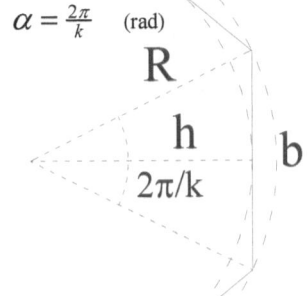

Mechanical Symmetry

Depending on the available data, we select one of the formulas contained in the following table.

Regular Polygons

	k	R	b	h
k	?	$\begin{bmatrix} b = 2R\sin\frac{\pi}{k} \\ h = R\cos\frac{\pi}{k} \end{bmatrix}$	$\begin{bmatrix} R = \dfrac{b}{2\sin\frac{\pi}{k}} \\ h = \dfrac{b}{2\tan\frac{\pi}{k}} \end{bmatrix}$	$\begin{bmatrix} b = 2h\tan\frac{\pi}{k} \\ R = \dfrac{h}{\cos\frac{\pi}{k}} \end{bmatrix}$
R	$\begin{bmatrix} b = 2R\sin\frac{\pi}{k} \\ h = R\cos\frac{\pi}{k} \end{bmatrix}$	?	$\begin{bmatrix} k = \dfrac{\pi}{\operatorname{asin}\frac{b}{2R}} \\ h = \dfrac{\sqrt{4R^2 - b^2}}{2} \end{bmatrix}$	$\begin{bmatrix} k = \dfrac{\pi}{\operatorname{acos}\frac{h}{R}} \\ b = 2\sqrt{R^2 - h^2} \end{bmatrix}$
b	$\begin{bmatrix} R = \dfrac{b}{2\sin\frac{\pi}{k}} \\ h = \dfrac{b}{2\tan\frac{\pi}{k}} \end{bmatrix}$	$\begin{bmatrix} k = \dfrac{\pi}{\operatorname{asin}\frac{b}{2R}} \\ h = \dfrac{\sqrt{4R^2 - b^2}}{2} \end{bmatrix}$	?	$\begin{bmatrix} k = \dfrac{\pi}{\operatorname{atan}\frac{b}{2h}} \\ R = \dfrac{\sqrt{4h^2 + b^2}}{2} \end{bmatrix}$
h	$\begin{bmatrix} b = 2h\tan\frac{\pi}{k} \\ R = \dfrac{h}{\cos\frac{\pi}{k}} \end{bmatrix}$	$\begin{bmatrix} k = \dfrac{\pi}{\operatorname{acos}\frac{h}{R}} \\ b = 2\sqrt{R^2 - h^2} \end{bmatrix}$	$\begin{bmatrix} k = \dfrac{\pi}{\operatorname{atan}\frac{b}{2h}} \\ R = \dfrac{\sqrt{4h^2 + b^2}}{2} \end{bmatrix}$	?

Eq. 34 Regular Polygons Resolution

Regular Polygon – b, h

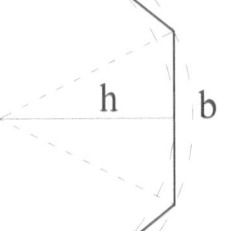

$$Area = k\frac{bh}{2}$$

$$I_k = k\frac{bh(12h^2+b^2)}{96} = k\left(\frac{bh^3}{8}+\frac{b^3h}{96}\right)$$

$$i_k = \sqrt{\frac{12h^2+b^2}{48}}$$

$$I_p = I_0 = k\frac{bh(12h^2+b^2)}{48} = k\left(\frac{bh^3}{4}+\frac{b^3h}{48}\right)$$

$$I_{kap} = k\frac{bh^3}{9}$$

$$\Delta I_k = \frac{1}{8}+\frac{3b^2}{32h^2}$$

Regular Polygon – R, k

$$Area = \frac{kR^2}{2}\sin\frac{2\pi}{k}$$

$$I_k = kR^4\frac{\sin\frac{2\pi}{k}\left(1+2\cos^2\frac{\pi}{k}\right)}{24}$$

$$i_k = R\sqrt{\frac{1+2\cos^2\frac{\pi}{k}}{12}}$$

$$I_p = I_0 = kR^4\frac{\sin\frac{2\pi}{k}\left(1+2\cos^2\frac{\pi}{k}\right)}{12}$$

$$I_{kap} = \frac{2kR^4\cos^3\frac{\pi}{k}\sin\frac{\pi}{k}}{9} \equiv kR^4\frac{\cos^2\frac{\pi}{k}\sin\frac{2\pi}{k}}{9}$$

$$\Delta I_k = \frac{1}{8}\left(1+3\tan^2\frac{\pi}{k}\right)$$

Regular Polygon – h , k

$$Area = kh^2 \tan\frac{\pi}{k}$$

$$I_k = kh^4 \frac{\tan^3\frac{\pi}{k} + 3\tan\frac{\pi}{k}}{12} = kh^4 \frac{\left(1+2\cos^2\frac{\pi}{k}\right)\sin\frac{\pi}{k}}{12\cos^3\frac{\pi}{k}}$$

$$i_k = h\sqrt{\frac{3+\tan^2\frac{\pi}{k}}{12}}$$

$$I_p = I_0 = kh^4 \frac{\tan^3\frac{\pi}{k} + 3\tan\frac{\pi}{k}}{6} \equiv kh^4 \frac{\left(1+2\cos^2\frac{\pi}{k}\right)\sin\frac{\pi}{k}}{6\cos^3\frac{\pi}{k}}$$

$$I_{kap} = \frac{8h^4 \tan\frac{\pi}{12}}{3}$$

$$\Delta I_k = \frac{1}{8}\left(1+3\tan^2\frac{\pi}{k}\right)$$

h=apo
$2\pi/k$

Regular Polygon – b , h

$$Area = \frac{kb^2}{4\tan\frac{\pi}{k}}$$

$$I_k = \frac{kb^4}{192\tan^3\frac{\pi}{k}}\left(3+\tan^2\frac{\pi}{k}\right) = kb^4 \frac{\left(\sin\frac{\pi}{k}\sin\frac{2\pi}{k} - 3\cos\frac{\pi}{k}\right)}{192\sin^3\frac{\pi}{k}}$$

$$i_k = b\sqrt{\frac{1+\dfrac{3}{\tan^2\frac{\pi}{k}}}{48}}$$

$$I_p = I_0 = \frac{kb^4}{96\tan^3\frac{\pi}{k}}\left(3+\tan^2\frac{\pi}{k}\right) \equiv kb^4 \frac{\left(\sin\frac{\pi}{k}\sin\frac{2\pi}{k} - 3\cos\frac{\pi}{k}\right)}{96\sin^3\frac{\pi}{k}}$$

$$I_{kap} = \frac{kb^4}{72\tan^3\frac{\pi}{k}}$$

$$\Delta I_k = \frac{1}{8}\left(1+3\tan^2\frac{\pi}{k}\right)$$

$2\pi/k$ b

Equilateral Triangle – b , h

$$Area = \frac{3}{2}bh$$

$$I_k = \frac{bh(12h^2 + b^2)}{32} \equiv \frac{3bh^3}{8} + \frac{b^3h}{32}$$

$$i_k = \sqrt{\frac{12h^2 + b^2}{48}}$$

$$I_p = I_0 = \frac{bh(12h^2 + b^2)}{16} \equiv \frac{3bh^3}{4} + \frac{b^3h}{16}$$

$$I_{kap} = k\frac{bh^3}{3}$$

$$\Delta I_k = 1.25$$

Equilateral Triangle – R

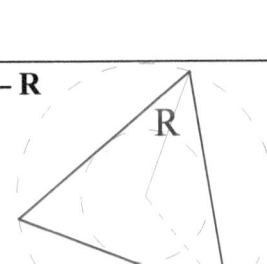

$$Area = \frac{\sqrt{27}}{4}R^2 \equiv 1.2990R^2$$

$$I_k = \frac{\sqrt{27}}{32}R^4 \equiv 0.16238R^4$$

$$i_k = \frac{R}{\sqrt{8}} = 0.35355R$$

$$I_p = I_0 = \frac{\sqrt{27}}{16}R^4 \equiv 0.32476R^4$$

$$I_{kap} = \frac{R^4}{8\sqrt{3}} \equiv 0.072169R^4$$

$$\Delta I_k = 1.25$$

Equilateral Triangle – h

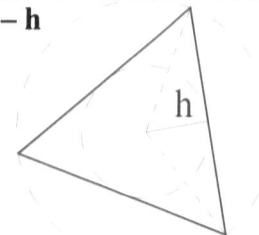

$Area = h^2\sqrt{27} \equiv 5.1962h^2$

$I_k = \dfrac{h^4\sqrt{27}}{2} = 2.598h^4$

$i_k = \dfrac{h}{\sqrt{2}} = 0.7071h$

$I_p = I_0 = h^4\sqrt{27} = 5.196h^4$

$I_{kap} = \dfrac{2h^4}{\sqrt{3}} \equiv 1.1547h^4$

$\Delta I_k = 1.25$

Equilateral Triangle – b

$Area = \dfrac{\sqrt{3}b^2}{4} \equiv 0.433b^2$

$I_k = \dfrac{b^4}{32\sqrt{3}} = 0.0180b^4$

$i_k = \dfrac{b}{\sqrt{24}} = 0.20412b$

$I_p = I_0 = \dfrac{b^4}{16\sqrt{3}} = 0.0361b^4$

$I_{kap} = \dfrac{b^4}{72\sqrt{3}} \equiv 0.00801875b^4$

$\Delta I_k = 1.25$

Square – b, h

$Area = 2bh$

$I_k = \dfrac{bh^3}{2} + \dfrac{b^3h}{24} = 0.5bh^3 + 0.041\widehat{6}b^3h$

$i_k = \dfrac{\sqrt{12h^2 + b^2}}{4\sqrt{3}} = 0.14434\sqrt{12h^2 + b^2}$

$I_p = I_0 = bh^3 + \dfrac{b^3h}{12} = 0.08\widehat{3}bh(12h^2 + b^2)$

$I_{kap} = \dfrac{4bh^3}{9} \equiv 0.\widehat{4}bh^3$

$\Delta I_k = \dfrac{1}{2}$

Square – R

$Area = 2R^2$

$I_k = \dfrac{R^4}{3} = 0.\widehat{3}R^4$

$i_k = \dfrac{R}{\sqrt{6}} = 0.40825R$

$I_p = I_0 = \dfrac{2R^4}{3} = 0.\widehat{6}R^4$

$I_{kap} = \dfrac{2R^4}{9} \equiv 0.\widehat{2}R^4$

$\Delta I_k = \dfrac{1}{2}$

Square – h

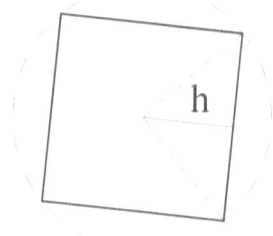

$Area = 4h^2$

$I_k = \dfrac{4h^4}{3} = 1.\widehat{3}h^4$

$i_k = \dfrac{h}{\sqrt{3}} = 0.57735h$

$I_p = I_0 = \dfrac{8h^4}{3} = 2.\widehat{6}h^4$

$I_{kap} = \dfrac{8h^4}{9} \equiv 0.\widehat{8}h^4$

$\Delta I_k = \dfrac{1}{2}$

Square – b

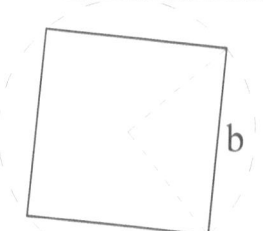

$Area = b^2$

$I_k = \dfrac{b^4}{12} = 0.08\widehat{3}b^4$

$i_k = \dfrac{b}{2\sqrt{3}} = 0.288675b$

$I_p = I_0 = \dfrac{b^4}{6} = 0.1\widehat{6}b^4$

$I_{kap} = \dfrac{b^4}{18} \equiv 0.0\widehat{5}b^4$

$\Delta I_k = \dfrac{1}{2}$

Pentagon – b, h

$$Area = \frac{5bh}{2}$$

$$I_k = \frac{5bh^3}{8} + \frac{5b^3h}{96} = 0.625bh^3 + 0.05208\widehat{3}b^3h$$

$$i_k = \frac{\sqrt{12h^2 + b^2}}{4\sqrt{3}} = 0.144337\sqrt{12h^2 + b^2}$$

$$I_p = I_0 = \frac{5bh(12h^2 + b^2)}{48} = 0.1041\widehat{6}bh(12h^2 + b^2)$$

$$I_{kap} = \frac{5bh^3}{9} \equiv 0.\widehat{5}bh^3$$

$$\Delta I_k = 0.3223$$

Pentagon – R

$$Area = 5\cos\tfrac{\pi}{5}\sin\tfrac{\pi}{5}R^2 \equiv 2.37764R^2$$

$$I_k = \frac{5}{24}\sin\tfrac{2\pi}{5}\left(1+2\cos^2\tfrac{\pi}{5}\right)R^4 = 0.45750R^4$$

$$i_k = R\sqrt{\frac{1+2\cos^2\tfrac{\pi}{5}}{12}} = 0.43865485619628R$$

$$I_p = I_0 = \frac{5}{12}\sin\tfrac{2\pi}{5}\left(1+2\cos^2\tfrac{\pi}{5}\right)R^4 = 0.9150R^4$$

$$I_{kap} = \frac{5}{9}\sin\tfrac{2\pi}{5}\cos^2\tfrac{\pi}{5}R^4 \equiv 0.34582R^4$$

$$\Delta I_k = 0.3223$$

Pentagon – h

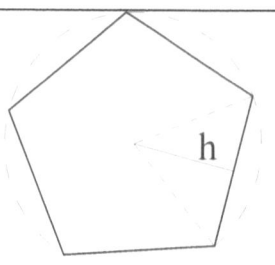

$$Area = 5\tan\tfrac{\pi}{5} h^2 \equiv 3.63271 h^2$$

$$I_k = \frac{5}{4}\left(\tan\tfrac{\pi}{5} + \frac{\tan^3\tfrac{\pi}{5}}{3}\right) h^4 = 1.067977 h^4$$

$$i_k = h\sqrt{\frac{3 + \tan^2\tfrac{\pi}{5}}{12}} = 0.54221 h$$

$$I_p = I_0 = \frac{5}{2}\left(\tan\tfrac{\pi}{5} + \frac{\tan^3\tfrac{\pi}{5}}{3}\right) h^4 = 2.135953 h^4$$

$$I_{kap} = \frac{10\tan\tfrac{\pi}{5}}{9} h^4 = 0.80727 h^4$$

$$\Delta I_k = 0.3223$$

Pentagon – b

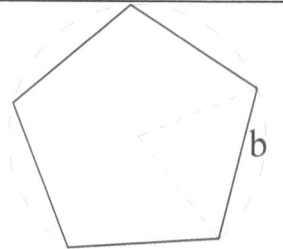

$$Area = \frac{5b^2}{4\tan\tfrac{\pi}{5}} \equiv 1.720477 b^2$$

$$I_k = \frac{5}{64}\left(\frac{1}{3\tan\tfrac{\pi}{5}} + \frac{1}{\tan^3\tfrac{\pi}{5}}\right) b^4 = 0.23955 b^4$$

$$i_k = b\sqrt{\frac{1}{48} + \frac{1}{16\tan^2\tfrac{\pi}{5}}} = 0.373142 b$$

$$I_p = I_0 = \frac{5}{32}\left(\frac{1}{3\tan\tfrac{\pi}{5}} + \frac{1}{\tan^3\tfrac{\pi}{5}}\right) b^4 = 0.4791 b^4$$

$$I_{kap} = \frac{5b^4}{72\tan^3\tfrac{\pi}{5}} \equiv 0.181$$

$$\Delta I_k = 0.3223$$

Hexagon – b , h

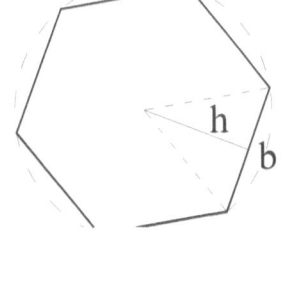

$Area = 3bh$

$I_k = \dfrac{3bh^3}{4} + \dfrac{b^3h}{16} = 0.75bh^3 + 0.0625b^3h$

$i_k = \dfrac{\sqrt{12h^2 + b^2}}{4\sqrt{3}} = 0.144337\sqrt{12h^2 + b^2}$

$I_p = I_0 = \dfrac{bh(12h^2 + b^2)}{8} = 0.125bh(12h^2 + b^2)$

$I_{kap} = \dfrac{2bh^3}{3} \equiv 0.\widehat{6}bh^3$

$\Delta I_k = 0.25$

Hexagon – R

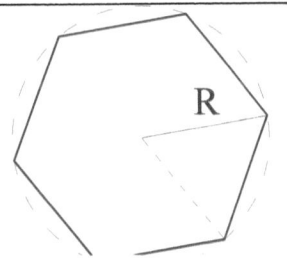

$Area = \dfrac{\sqrt{27}}{2}R^2 \equiv 2.5981R^2$

$I_k = \dfrac{5\sqrt{3}R^4}{16} = 0.54127R^4$

$i_k = R\sqrt{\dfrac{5}{24}} = 0.45644R$

$I_p = I_0 = \dfrac{5\sqrt{3}R^4}{8} = 1.08253R^4$

$I_{kap} = \dfrac{\sqrt{3}}{4}R^4 \equiv 0.43301R^4$

$\Delta I_k = 0.25$

Hexagon – h

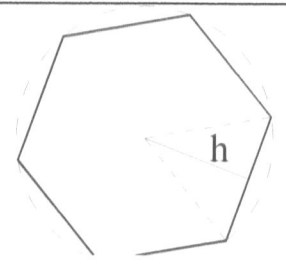

$Area = 2\sqrt{3}h^2 \equiv 3.4641h^2$

$I_k = \dfrac{5}{\sqrt{27}}h^4 = 0.96225h^4$

$i_k = h\sqrt{\dfrac{5}{18}} = 0.527h$

$I_p = I_0 = \dfrac{10}{\sqrt{27}}h^4 = 1.9245h^4$

$I_{kap} = \dfrac{4}{\sqrt{27}}h^4 \equiv 0.7698h^4$

$\Delta I_k = 0.25$

Hexagon – b

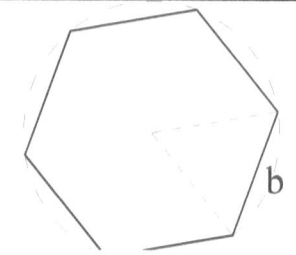

$Area = \dfrac{\sqrt{27}}{2}b^2 \equiv 2.5981b^2$

$I_k = \dfrac{5\sqrt{3}b^4}{16} = 0.54127b^4$

$i_k = b\sqrt{\dfrac{5}{24}} = 0.45644b$

$I_p = I_0 = \dfrac{5\sqrt{3}b^4}{8} = 1.08253b^4$

$I_{kap} = \dfrac{\sqrt{3}b^4}{4} \equiv 0.43301b^4$

$\Delta I_k = 0.25$

Heptagon

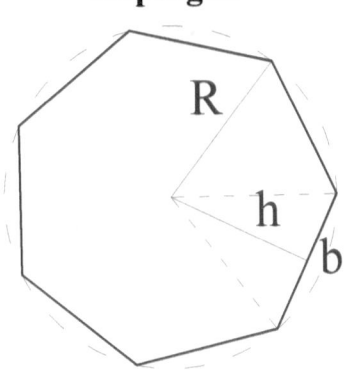

Area

$3.5bh$

$2.736410188638105R^2$

$3.3710223316527h^2$

$3.633912444001589b^2$

I_k

$0.875bh^3 + 0.07291\widehat{6}b^3h$

$0.5982453519662R^4$

$0.90790455421028h^4$

$1.055032537935482b^4$

i_k

$0.14433756729741\sqrt{12h^2 + b^2}$

$0.46757261484704R$

$0.51896644990144h$

$0.53882246866176b$

I_{kap}

$0.\widehat{7}bh^3$

$0.49361489277292R^4$

$0.7491160737006h^4$

$0.87051202549818b^4$

$I_p = I_0$

$0.1458\widehat{3}bh(12h^2 + b^2)$

$1.1964907039324R^4$

$1.815809108420554h^4$

$2.110065075870964b^4$

ΔI_k

0.21196779255486

$I_{xy} = 0 \Rightarrow I_u = I_v = I_x = I_y = I_k$

$i_{xy} = 0 \Rightarrow i_u = i_v = i_x = i_y = i_k$

Octagon

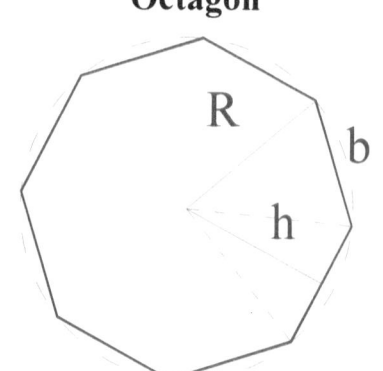

Area

$4bh$

$2.82842712474619 R^2$

$3.31370849898476 h^2$

$4.82842712474619 b^2$

I_k

$bh^3 + 0.08\widehat{3} b^3 h$

$0.6380711874577 R^4$

$0.87580566598984 h^4$

$1.85947570824873 b^4$

i_k

$0.14433756729741 \sqrt{12h^2 + b^2}$

$0.47496550586916 R$

$0.51409895896071 h$

$0.62057233956242 b$

I_{kap}

$0.\widehat{8} bh^3$

$0.53649190274958 R^4$

$0.73637966644106 h^4$

$1.563451979096164 b^4$

$I_p = I_0$

$0.1\widehat{6} bh(12h^2 + b^2)$

$1.276142374915396 R^4$

$1.75161133197968 h^4$

$3.71895141649746 b^4$

ΔI_k

0.18933982822018

$I_{xy} = 0 \Rightarrow I_u = I_v = I_x = I_y = I_k$

$i_{xy} = 0 \Rightarrow i_u = i_v = i_x = i_y = i_k$

Dodecagon

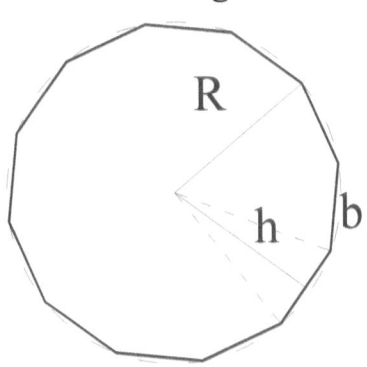

Area

$6bh$

$3R^2$

$3.2153903091734721h^2$

$11.19615242270663b^2$

I_k	i_k
$1.5bh^3 + 0.125b^3h$	$0.14433756729741\sqrt{12h^2+b^2}$
$0.71650635094611R^4$	$0.48870793968931R$
$0.82308546376021h^4$	$0.50594768913763h$
$9.979646071760525b^4$	$0.94411124091685b$

I_{kap}	$I_p = I_0$
$1.\widehat{3}bh^3$	$0.25bh(12h^2+b^2)$
$0.62200846792815R^4$	$1.433012701892219R^4$
$0.71453117981633h^4$	$1.646170927520417h^4$
$8.663460352255527b^4$	$19.95929214352104b^4$

ΔI_k	
0.15192378864668	$I_{xy}=0 \Rightarrow I_u = I_v = I_x = I_y = I_k$
	$i_{xy}=0 \Rightarrow i_u = i_v = i_x = i_y = i_k$

Regular Polygons – R

$$Area = R^2 Area_k \quad I_k = R^4 MoI_k \quad i_k = R \cdot RadGir_k$$
$$I_p = R^4 PolMoI_k \quad I_{kap} = R^4 ApMoI_k$$

k	$Area_k$	MoI_k	$RadGir_k$	$PolMoI_k$	$ApMoI_k$	ApMol Err
3	1,299	0,16238	0,35355	0,32476	0,072169	125,0%
4	2	0,33333	0,40825	0,66667	0,22222	50,0%
5	2,3776	0,4575	0,43865	0,915	0,34582	32,3%
6	2,5981	0,54127	0,45644	1,0825	0,43301	25,0%
7	2,7364	0,59825	0,46757	1,1965	0,49361	21,2%
8	2,8284	0,63807	0,47497	1,2761	0,53649	18,9%
9	2,8925	0,66674	0,48011	1,3335	0,5676	17,5%
10	2,9389	0,68796	0,48382	1,3759	0,59073	16,5%
12	3	0,71651	0,48871	1,433	0,62201	15,2%
13	3,0207	0,72634	0,49036	1,4527	0,63282	14,8%
14	3,0372	0,73423	0,49168	1,4685	0,64151	14,5%
15	3,0505	0,74065	0,49274	1,4813	0,64859	14,2%
16	3,0615	0,74595	0,49362	1,4919	0,65443	14,0%
17	3,0706	0,75036	0,49434	1,5007	0,65931	13,8%
18	3,0782	0,75408	0,49495	1,5082	0,66341	13,7%
19	3,0846	0,75723	0,49546	1,5145	0,66691	13,5%
20	3,0902	0,75994	0,4959	1,5199	0,6699	13,4%
30	3,1187	0,77399	0,49818	1,548	0,68547	12,9%
40	3,1287	0,77896	0,49897	1,5579	0,69098	12,7%
50	3,1333	0,78127	0,49934	1,5625	0,69355	12,6%
60	3,1359	0,78253	0,49954	1,5651	0,69495	12,6%
70	3,1374	0,78329	0,49966	1,5666	0,69579	12,6%
80	3,1384	0,78378	0,49974	1,5676	0,69634	12,6%
90	3,139	0,78412	0,4998	1,5682	0,69672	12,5%
100	3,1395	0,78437	0,49984	1,5687	0,69698	12,5%
200	3,1411	0,78514	0,49996	1,5703	0,69784	12,5%
300	3,1414	0,78528	0,49998	1,5706	0,698	12,5%
400	3,1415	0,78533	0,49999	1,5707	0,69806	12,5%
500	3,1415	0,78536	0,49999	1,5707	0,69809	12,5%
32	3,1214	0,77536	0,4984	1,5507	0,68699	12,9%
64	3,1365	0,78288	0,4996	1,5658	0,69533	12,6%
128	3,1403	0,78477	0,4999	1,5695	0,69743	12,5%
256	3,1413	0,78524	0,49997	1,5705	0,69796	12,5%
512	3,1415	0,78536	0,49999	1,5707	0,69809	12,5%
1000	3,1416	0,78539	0,5	1,5708	0,69812	12,5%

Regular Polygons – h

$$Area = h^2 Area_k \quad I_k = h^4 MoI_k \quad i_k = h \cdot RadGir_k$$

$$I_p = h^4 PolMoI_k \quad I_{kap} = h^4 ApMoI_k$$

k	Areak	MoIk	RadGirk	PolMoIk	ApMoIk	ApMoI Err
3	5,1962	2,5981	0,70711	5,1962	1,1547	125,0%
4	4	1,3333	0,57735	2,6667	0,88889	50,0%
5	3,6327	1,068	0,54221	2,136	0,80727	32,3%
6	3,4641	0,96225	0,52705	1,9245	0,7698	25,0%
7	3,371	0,9079	0,51897	1,8158	0,74912	21,2%
8	3,3137	0,87581	0,5141	1,7516	0,73638	18,9%
9	3,2757	0,8551	0,51092	1,7102	0,72794	17,5%
10	3,2492	0,84088	0,50872	1,6818	0,72204	16,5%
12	3,2154	0,82309	0,50595	1,6462	0,71453	15,2%
13	3,2042	0,81727	0,50504	1,6345	0,71205	14,8%
14	3,1954	0,81272	0,50432	1,6254	0,71009	14,5%
15	3,1883	0,80909	0,50375	1,6182	0,70852	14,2%
16	3,1826	0,80614	0,50329	1,6123	0,70724	14,0%
17	3,1779	0,80372	0,5029	1,6074	0,70619	13,8%
18	3,1739	0,80169	0,50258	1,6034	0,70531	13,7%
19	3,1705	0,79999	0,50232	1,6	0,70456	13,5%
20	3,1677	0,79854	0,50209	1,5971	0,70393	13,4%
30	3,1531	0,79118	0,50092	1,5824	0,70069	12,9%
40	3,1481	0,78864	0,50052	1,5773	0,69957	12,7%
50	3,1457	0,78747	0,50033	1,5749	0,69905	12,6%
60	3,1445	0,78684	0,50023	1,5737	0,69877	12,6%
70	3,1437	0,78645	0,50017	1,5729	0,6986	12,6%
80	3,1432	0,78621	0,50013	1,5724	0,69849	12,6%
90	3,1429	0,78604	0,5001	1,5721	0,69842	12,5%
100	3,1426	0,78592	0,50008	1,5718	0,69836	12,5%
200	3,1419	0,78553	0,50002	1,5711	0,69819	12,5%
300	3,1417	0,78546	0,50001	1,5709	0,69816	12,5%
400	3,1417	0,78543	0,50001	1,5709	0,69815	12,5%
500	3,1416	0,78542	0,5	1,5708	0,69814	12,5%
32	3,1517	0,79048	0,50081	1,581	0,70038	12,9%
64	3,1441	0,78666	0,5002	1,5733	0,69869	12,6%
128	3,1422	0,78571	0,50005	1,5714	0,69827	12,5%
256	3,1418	0,78548	0,50001	1,571	0,69817	12,5%
512	3,1416	0,78542	0,5	1,5708	0,69814	12,5%
1000	3,1416	0,7854	0,5	1,5708	0,69813	12,5%

Regular Polygons – b

$$Area = b^2 Area_k \quad I_k = b^4 MoI_k \quad i_k = b \cdot RadGir_k$$
$$I_p = b^4 PolMoI_k \quad I_{kap} = b^4 ApMoI_k$$

k	$Area_k$	MoI_k	$RadGir_k$	$PolMoI_k$	$ApMoI_k$	ApMoI Err
3	0,43301	0,018042	0,20412	0,036084	0,0080188	125,0%
4	1	0,083333	0,28868	0,16667	0,055556	50,0%
5	1,7205	0,23955	0,37314	0,4791	0,18107	32,3%
6	2,5981	0,54127	0,45644	1,0825	0,43301	25,0%
7	3,6339	1,055	0,53882	2,1101	0,87051	21,2%
8	4,8284	1,8595	0,62057	3,719	1,5635	18,9%
9	6,1818	3,0453	0,70187	6,0906	2,5925	17,5%
10	7,6942	4,7153	0,78284	9,4307	4,0489	16,5%
12	11,196	9,9796	0,94411	19,959	8,6635	15,2%
13	13,186	13,84	1,0245	27,68	12,058	14,8%
14	15,335	18,717	1,1048	37,433	16,353	14,5%
15	17,642	24,773	1,185	49,546	21,694	14,2%
16	20,109	32,184	1,2651	64,369	28,236	14,0%
17	22,735	41,138	1,3451	82,276	36,146	13,8%
18	25,521	51,834	1,4251	103,67	45,602	13,7%
19	28,465	64,483	1,5051	128,97	56,791	13,5%
20	31,569	79,31	1,585	158,62	69,913	13,4%
30	71,358	405,21	2,383	810,41	358,86	12,9%
40	127,06	1284,8	3,1798	2569,5	1139,7	12,7%
50	198,68	3141,3	3,9763	6282,6	2788,6	12,6%
60	286,22	6519	4,7725	13038	5789,4	12,6%
70	389,67	12083	5,5686	24166	10733	12,6%
80	509,03	20620	6,3646	41240	18319	12,6%
90	644,32	33036	7,1605	66072	29353	12,5%
100	795,51	50360	7,9564	1,01E+05	44750	12,5%
200	3182,8	8,06E+05	15,915	1,61E+06	7,17E+05	12,5%
300	7161,7	4,08E+06	23,873	8,16E+06	3,63E+06	12,5%
400	12732	1,29E+07	31,831	2,58E+07	1,15E+07	12,5%
500	19894	3,15E+07	39,788	6,30E+07	2,80E+07	12,5%
32	81,225	525,02	2,5424	1050	465,18	12,9%
64	325,69	8441	5,0909	16882	7497,1	12,6%
128	1303,5	1,35E+05	10,185	2,70E+05	1,20E+05	12,5%
256	5214,9	2,16E+06	20,371	4,33E+06	1,92E+06	12,5%
512	20860	3,46E+07	40,743	6,93E+07	3,08E+07	12,5%
1000	79418	5,02E+08	79,498	1,00E+09	4,46E+08	12,5%

6.3 Regular Polygonal Tubes

Mechanical Symmetry

Regular Polygonal Tube – h_2, b_2, t

Area $= \dfrac{b_2 t(2h_2 - t)}{2h_2} k$

$I_k = \dfrac{b_2\left(12h_2^2 + b_2^2\right) t(2h_2 - t)\left(t^2 - 2h_2 t + 2h_2^2\right)}{96 h_2^3} k$

$I_p = I_0 = \dfrac{b_2\left(12h_2^2 + b_2^2\right) t(2h_2 - t)\left(t^2 - 2h_2 t + 2h_2^2\right)}{48 h_2^3} k$

$i_k = \dfrac{\sqrt{\left(t^2 - 2h_2 t + 2h_2^2\right)\left(12h_2^2 + b_2^2\right)}}{h_2 \, 4\sqrt{3}}$

$I_{kap} = \dfrac{b_2 k t\left(t^2 - 3h_2 t + 3h_2^2\right)^2}{9 h_2 (2h_2 - t)}$

$\Delta I_k = \dfrac{\left(4h_2^2 + 3b_2^2\right)t^4 + \left(-24h_2^3 - 18 b_2^2 h_2\right)t^3 + \left(24h_2^4 + 42 b_2^2 h_2^2\right)t^2 - 48 b_2^2 h_2^3 t + 24 b_2^2 h_2^4}{32 h_2^2 t^4 - 192 h_2^3 t^3 + 480 h_2^4 t^2 - 576 h_2^5 t + 288 h_2^6}$

Regular Polygonal Tube – b_2, t

Area $= t\left(b_2 - t \cdot \tan\tfrac{\pi}{k}\right) k$

$I_k = \dfrac{t\left(\tan^2\tfrac{\pi}{k} + 3\right)\left(b_2 - t\tan\tfrac{\pi}{k}\right)\left(2t^2\tan^2\tfrac{\pi}{k} - 2tb_2\tan\tfrac{\pi}{k} + b_2^2\right)}{24\tan^2\tfrac{\pi}{k}} k$

$I_p = I_0 = \dfrac{t\left(\tan^2\tfrac{\pi}{k} + 3\right)\left(b_2 - t\tan\tfrac{\pi}{k}\right)\left(2t^2\tan^2\tfrac{\pi}{k} - 2tb_2\tan\tfrac{\pi}{k} + b_2^2\right)}{12\tan^2\tfrac{\pi}{k}} k$

$i_k = \sqrt{\dfrac{\left(\tan^2\tfrac{\pi}{k} + 3\right)\left(2t^2\tan^2\tfrac{\pi}{k} - 2tb_2\tan\tfrac{\pi}{k} + b_2^2\right)}{24\tan^2\tfrac{\pi}{k}}}$

$I_{kap} = \dfrac{t\left(4t^2\tan^2\tfrac{\pi}{k} - 6 b_2 t \cdot \tan\tfrac{\pi}{k} + 3b_2^2\right)^2}{72\tan^2\tfrac{\pi}{k}\left(b_2 - t\cdot\tan\tfrac{\pi}{k}\right)} k$

$\Delta I_k = \dfrac{\tan^2\tfrac{\pi}{k}\left(6 t^4 \tan^4\tfrac{\pi}{k} + 2 t^4 \tan^2\tfrac{\pi}{k} - 18 b_2 t^3 \tan^3\tfrac{\pi}{k} - 6 b_2 t^3 \tan\tfrac{\pi}{k} + 21 b_2^2 t^2 \tan^2\tfrac{\pi}{k} + 3 b_2^2 t^2 - 12 t b_2^3 \tan\tfrac{\pi}{k} + 3 b_2^4\right)}{\left(4 t^2 \tan^2\tfrac{\pi}{k} - 6 t b_2 \tan\tfrac{\pi}{k} + 3 b_2^2\right)^2}$

Regular Polygonal Tube – h2 , t

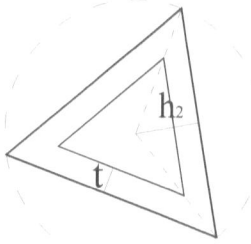

Area = $\left(\tan\frac{\pi}{k}\right)t(2h_2 - t)k$

$$I_k = \frac{\tan\frac{\pi}{k}\left(\tan^2\frac{\pi}{k}+3\right)t(2h_2-t)\left[(2h_2-t)^2 - t^2\right]}{24}k$$

$$Ip = I_0 = \frac{\tan\frac{\pi}{k}\left(\tan^2\frac{\pi}{k}+3\right)t(2h_2-t)(t^2 - 2h_2t + 2h_2^2)}{6}k$$

$$i_k = \sqrt{\frac{\left(\tan^2\frac{\pi}{k}+3\right)(t^2 - 2h_2t + 2h_2^2)}{12}}$$

$$I_{kap} = \frac{2\left(\tan\frac{\pi}{k}\right)t\left(t^2 - 3h_2t + 3h_2^2\right)^2}{9(2h_2 - t)}k$$

$$\Delta I_k = \frac{2\sin^2\frac{\pi}{k}t^4 + t^4 - 12h_2t^3\sin^2\frac{\pi}{k} - 6h_2t^3 + 36h_2^2t^2\sin^2\frac{\pi}{k} + 6h_2^2t^2 - 48th_2^3\sin^2\frac{\pi}{k} + 24h_2^4\sin^2\frac{\pi}{k}}{8\cos^2\frac{\pi}{k}(t^2 - 3h_2t + 3h_2^2)^2}$$

Regular Polygonal Tube – R₂ , t

Area = $\tan\frac{\pi}{k}t\left(2R_2\cos\frac{\pi}{k} - t\right)k$

$$I_k = \frac{\tan\frac{\pi}{k}\left(1 + 2\cos^2\frac{\pi}{k}\right)t\left(2R_2\cos\frac{\pi}{k} - t\right)\left[\left(2R_2\cos\frac{\pi}{k} - t\right)^2 - t^2\right]}{24\cos^2\frac{\pi}{k}}k$$

$$I_p = I_0 = \frac{\tan\frac{\pi}{k}\left(1 + 2\cos^2\frac{\pi}{k}\right)t\left(2R_2\cos\frac{\pi}{k} - t\right)\left(2R_2^2\cos^2\frac{\pi}{k} - 2tR_2\cos\frac{\pi}{k} + t^2\right)}{6\cos^2\frac{\pi}{k}}k$$

$$i_k = \sqrt{\frac{\left(1 + 2\cos^2\frac{\pi}{k}\right)\left(2R_2^2\cos^2\frac{\pi}{k} - 2tR_2\cos\frac{\pi}{k} + t^2\right)}{12\cos^2\frac{\pi}{k}}}$$

$$\Delta I_k = -\frac{24\left[\left(\cos^6\frac{\pi}{k} - \cos^4\frac{\pi}{k}\right)R_2^4 + \left(\cos^3\frac{\pi}{k} - \cos^5\frac{\pi}{k}\right)2tR_2^3\right] + \left(6\cos^4\frac{\pi}{k} - 7\cos^2\frac{\pi}{k}\right)6t^2R_2^2 + \left(6\cos\frac{\pi}{k} - 4\cos^3\frac{\pi}{k}\right)3t^3R_2 + \left(2\cos^2\frac{\pi}{k} - 3\right)t^4}{8\cos^2\frac{\pi}{k}\left(3R_2^2\cos^2\frac{\pi}{k} - 3tR_2\cos\frac{\pi}{k} + t^2\right)^2}$$

$$I_{kap} = \frac{2\tan\frac{\pi}{k}t\left(3R_2^2\cos^2\frac{\pi}{k} - 3tR_2\cos\frac{\pi}{k} + t^2\right)^2}{9\left(2R_2\cos\frac{\pi}{k} - t\right)}k$$

Regular Polygonal Tube – h_2, h_1

Area = $\tan\frac{\pi}{k}(h_2 - h_1)(h_2 + h_1)k$

$I_k = \dfrac{(h_2 - h_1)(h_2 + h_1)(h_2^2 + h_1^2)\tan\frac{\pi}{k}(\tan^2\frac{\pi}{k} + 3)k}{12}$

$I_p = I_0 = \dfrac{(h_2^4 - h_1^4)\tan^3\frac{\pi}{k} + (3h_2^4 - 3h_1^4)\tan\frac{\pi}{k}}{6}k$

$i_k = \sqrt{\dfrac{(\tan^2\frac{\pi}{k} + 3)(h_2^2 + h_1^2)}{12}}$

$I_{kap} = \dfrac{2(h_2 - h_1)(h_2^2 + h_1 h_2 + h_1^2)^2 \tan\frac{\pi}{k}}{9(h_2 + h_1)}k$

$\Delta I_k = \dfrac{3h_2^4 + 6h_1 h_2^3 + 6h_1^2 h_2^2 + 6h_1^3 h_2 + 3h_1^4 - (2h_2^4 + 4h_1 h_2^3 + 12h_1^2 h_2^2 + 4h_1^3 h_2 + 2h_1^4)\cos^2\frac{\pi}{k}}{(8h_2^4 + 16h_1 h_2^3 + 24h_1^2 h_2^2 + 16h_1^3 h_2 + 8h_1^4)\cos^2\frac{\pi}{k}}$

Regular Polygonal Tube - R_1, R_2

Area = $\dfrac{\sin\frac{2\pi}{k}(R_2 - R_1)(R_2 + R_1)}{2}k$

$I_k = \dfrac{\sin\frac{2\pi}{k}(1 + 2\cos^2\frac{\pi}{k})(R_2 - R_1)(R_2 + R_1)(R_2^2 + R_1^2)}{24}k$

$I_p = I_0 = \dfrac{\sin\frac{2\pi}{k}(1 + 2\cos^2\frac{\pi}{k})(R_2 - R_1)(R_2 + R_1)(R_2^2 + R_1^2)}{12}k$

$i_k = \sqrt{\dfrac{(R_2^2 + R_1^2)(1 + 2\cos^2\frac{\pi}{k})}{12}}$

$\Delta I_k = \dfrac{(2 - \cos\frac{2\pi}{k})(R_2^4 + R_1^4) + (4 - 2\cos\frac{2\pi}{k})(R_1 R_2^3 + R_1^3 R_2) - 6\cos\frac{2\pi}{k} R_1^2 R_2^2}{8\cos^2\frac{\pi}{k}(R_2^2 + R_1 R_2 + R_1^2)^2}$

$I_{kap} = \dfrac{\sin\frac{2\pi}{k}\cos^2\frac{\pi}{k}(R_2 - R_1)(R_2^2 + R_1 R_2 + R_1^2)^2}{9(R_2 + R_1)}k$

Regular Hexagonal Tube

$$\text{Área} = \begin{cases} t(1.73205b_2 - t)3.4641 \\ t(2h_2 - t)3.4641 \\ t(1.73205R_2 - t)3.4641 \end{cases}$$

$$I_k = \begin{cases} t(6.92820b_2 t^2 - 9b_2^2 t + 5.19615b_2^3 - 2t^3)0.481125 \\ t(2h_2 - t)(t^2 - 2h_2 t + 2h_2^2)0.9623 \\ t(6.92820R_2 t^2 - 9R_2^2 t + 5.19615R_2^3 - 2t^3)0.481125 \end{cases}$$

$$I_P = I_0 = \begin{cases} t(6.9282b_2 t^2 - 9b_2^2 t + 5.19615b_2^3 - 2t^3)0.9623 \\ t(2h_2 - t)(t^2 - 2h_2 t + 2h_2^2)1.9245 \\ t(6.9282R_2 t^2 - 9R_2^2 t + 5.19615R_2^3 - 2t^3)0.9623 \end{cases}$$

$$i_k = \begin{cases} 0.3727\sqrt{\dfrac{2t^3 - 6.9282b_2 t^2 + 9b_2^2 t - 5.19615b_2^3}{t - 1.73205b_2}} \\ 0.527\sqrt{t^2 - 2h_2 t + 2h_2^2} \\ 0.3727\sqrt{\dfrac{2t^3 - 6.9282R_2 t^2 + 9R_2^2 t - 5.19615R_2^3}{t - 1.73205R_2}} \end{cases}$$

$$I_{kap} = \begin{cases} \dfrac{0.04811t(10.3923b_2 t - 9b_2^2 - 4t^2)^2}{1.73205b_2 - t} \\ \dfrac{0.7698t(t^2 - 3h_2 t + 3h_2^2)^2}{2h_2 - t} \end{cases}$$

$$\Delta I_k = \begin{cases} \dfrac{21(1.7b_2 - t)\left(\dfrac{0.05t(4t^2 - 10.b_2 t + 9b_2^2)^2}{t - 1.7b_2} - 0.5t(2t^3 - 6.9b_2 t^2 + 9b_2^2 t - 5.2b_2^3)\right)}{t(4t^2 - 10b_2 t + 9b_2^2)^2} \\ \dfrac{5656704 t^4 - 33940224 h_2 t^3 + 56567041 h_2^2 t^2 - 45253634 h_2^3 t + 22626817 h_2^4}{22626815(t^2 - 3h_2 t + 3h_2^2)^2} \end{cases}$$

Regular Hexagonal Tube

$$\acute{A}rea = \begin{cases} 3.4641\left(h_2^2 - h_1^2\right) \\ 2.59808\left(R_2^2 - R_1^2\right) \\ 2.59808\left(b_2^2 - b_1^2\right) \end{cases}$$

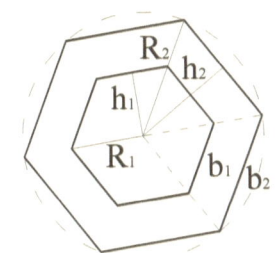

$$I_P = I_0 = \begin{cases} 1.9245\left(h_2^4 - h_1^4\right) \\ 1.08253\left(R_2^4 - R_1^4\right) \\ 1.08253\left(b_2^4 - b_1^4\right) \end{cases} \qquad I_k = \begin{cases} 0.9623\left(h_2^4 - h_1^4\right) \\ 0.5413\left(R_2^4 - R_1^4\right) \\ 0.5413\left(b_2^4 - b_1^4\right) \end{cases}$$

$$I_{kap} = \begin{cases} \dfrac{0.7698\left(h_2 - h_1\right)\left(h_2^2 + h_1 h_2 + h_1^2\right)^2}{h_2 + h_1} \\ \\ \dfrac{0.433\left(R_2 - R_1\right)\left(R_2^2 + R_1 R_2 + R_1^2\right)^2}{R_2 + R_1} \end{cases} \qquad i_k = \begin{cases} 0.527\sqrt{h_2^2 + h_1^2} \\ 0.4564\sqrt{R_2^2 + R_1^2} \\ 0.4564\sqrt{b_2^2 + b_1^2} \end{cases}$$

$$\Delta I_k = \begin{cases} \dfrac{5656704 h_2^4 + 11313408 h_1 h_2^3 - 11313407 h_1^2 h_2^2 + 11313408 h_1^3 h_2 + 5656704 h_1^4}{22626815\left(h_2^2 + h_1 h_2 + h_1^2\right)^2} \\ \\ \dfrac{5307335 R_2^4 + 10614670 R_1 R_2^3 - 10614671 R_1^2 R_2^2 + 10614670 R_1^3 R_2 + 5307335 R_1^4}{21229341\left(R_2^2 + R_1 R_2 + R_1^2\right)^2} \end{cases}$$

Regular Octagonal Tube

$$\acute{A}rea = \begin{cases} 8(b_2 - 0.4142t)t \\ 3.31371(2h_2 - t)t \\ 3.31371(1.84776 R_2 - t)t \end{cases}$$

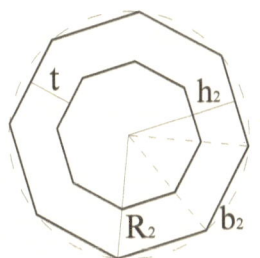

$$I_k = \begin{cases} 6.16176(b_2 - 0.4142t)t(0.3431t^2 - 0.8284b_2 t + b_2^2) \\ 0.8758(2h_2 - t)t(t^2 - 2h2t + 2h2^2) \\ 0.8758(1.84776R_2 - t)t(1.70711R_2^2 - 1.84776t R_2 + t^2) \end{cases}$$

$$I_P = I_0 = \begin{cases} 12.3235(b_2 - 0.4142t)t(0.3431t^2 - 0.8284b2t + b2^2) \\ 1.75161(2h_2 - t)t(t^2 - 2h2t + 2h2^2) \\ 1.75161(1.84776R_2 - t)t(1.70711R_2^2 - 1.84776t R_2 + t^2) \end{cases}$$

$$i_k = \begin{cases} 0.8776\sqrt{0.3431t^2 - 0.8284b_2 t + b_2^2} \\ 0.5141\sqrt{t^2 - 2h_2 t + 2h_2^2} \\ 0.5141\sqrt{1.70711R_2^2 - 1.84776t R_2 + t^2} \end{cases}$$

$$I_{kap} = \begin{cases} \dfrac{0.6476t(0.6863t^2 - 2.48528b_2 t + 3b_2^2)^2}{b_2 - 0.4142t} \\ \dfrac{0.7364t(t^2 - 3h_2 t + 3h_2^2)^2}{2h_2 - t} \\ \dfrac{0.7364t(2.56066R_2^2 - 2.77164t R_2 + t^2)^2}{1.84776R_2 - t} \end{cases}$$

$$\Delta I_k = \begin{cases} \dfrac{3(9t^4 - 454b_2 t^3 + 1375b_2^2 t^2 - 2300b_2^3 t + 2000b_2^4)}{50(7t^2 - 25b_2 t + 30b_2^2)^2} \\ \dfrac{92499611t^4 - 55499766h_2 t^3 + 80645697h_2^2 t^2 - 50291862h_2^3 t + 251459311h_2^4}{48853757(t^2 - 3h_2 t + 3h_2^2)^2} \\ \dfrac{3(9t^4 - 454b_2 t^3 + 1375b_2^2 t^2 - 2300b_2^3 t + 2000b_2^4)}{50(7t^2 - 25b_2 t + 30b_2^2)^2} \end{cases}$$

Regular Octagonal Tube

$$\acute{A}rea = \begin{cases} 3.31371\left(h_2^2 - h_1^2\right) \\ 2.82843\left(R_2^2 - R_1^2\right) \\ 4.82843\left(b_2^2 - b_1^2\right) \end{cases}$$

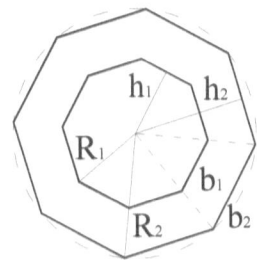

$$I_P = I_0 = \begin{cases} 1.75161\left(h_2^4 - h_1^4\right) \\ 1.27614\left(R_2^4 - R_1^4\right) \\ 3.71895\left(b_2^4 - b_1^4\right) \end{cases} \qquad I_k = \begin{cases} 0.8758\left(h_2^4 - h_1^4\right) \\ 0.6381\left(R_2^4 - R_1^4\right) \\ 1.85948\left(b_2^4 - b_1^4\right) \end{cases}$$

$$I_{kap} = \begin{cases} \dfrac{0.7364\left(h_2 - h_1\right)\left(h_2^2 + h_1 h_2 + h_1^2\right)^2}{h_2 + h_1} \\ \dfrac{0.5365\left(R_2 - R_1\right)\left(R_2^2 + R_1 R_2 + R_1^2\right)^2}{R_2 + R_1} \\ \dfrac{1.56345\left(b_2 - b_1\right)\left(b_2^2 + b_1 b_2 + b_1^2\right)^2}{b_2 + b_1} \end{cases} \qquad i_k = \begin{cases} 0.5141\sqrt{h_2^2 + h_1^2} \\ 0.475\sqrt{R_2^2 + R_1^2} \\ 0.6206\sqrt{b_2^2 + b_1^2} \end{cases}$$

$$\Delta I_k = \begin{cases} \dfrac{9249961 h_2^4 + 18499922 h_1 h_2^3 - 30353835 h_1^2 h_2^2 + 18499922 h_1^3 h_2 + 9249961 h_1^2}{48853757\left(h_2^2 + h_1 h_2 + h_1^2\right)^2} \\ \dfrac{19193 R_2^4 + 38386 R_1 R_2^3 - 62982 R_1^2 R_2^2 + 38386 R_1^3 R_2 + 19193 R_1^4}{101368\left(R_2^2 + R_1 R_2 + R_1^2\right)^2} \\ \dfrac{17185361 b_2^4 + 34370722 b_1 b_2^3 - 563939113 b_1^2 b_2^2 + 34370722 b_1^3 b2 + 17185361 b_1^4}{90764635\left(b_2^2 + b_1 b_2 + b_1^2\right)^2} \end{cases}$$

Mechanical Symmetry

Regular Polygonal Tubes – t, h_2

$$Area = Area_k \left(2h_2 - t\right)t$$

$$I_k = MoI_k\left(2h_2 - t\right)t\left(t^2 - 2h2t + 2h2^2\right) \quad i_k = RadGir_k\sqrt{t^2 - 2h_2 t + 2h_2^2}$$

$$I_p = PolMoI_k\left(2h_2 - t\right)t\left(t^2 - 2h2t + 2h2^2\right) \quad I_{kap} = ApMoI_k \frac{t\left(t^2 - 3h_2 t + 3h_2^2\right)^2}{2h_2 - t}$$

k	$Area_k$	MoI_k	$RadGir_k$	$PolMoI_k$	$ApMoI_k$
3	5,1962	2,5981	0,70711	5,1962	1,1547
4	4	1,3333	0,57735	2,6667	0,88889
5	3,6327	1,068	0,54221	2,136	0,80727
6	3,4641	0,96225	0,52705	1,9245	0,7698
7	3,371	0,9079	0,51897	1,8158	0,74912
8	3,3137	0,87581	0,5141	1,7516	0,73638
9	3,2757	0,8551	0,51092	1,7102	0,72794
10	3,2492	0,84088	0,50872	1,6818	0,72204
12	3,2154	0,82309	0,50595	1,6462	0,71453
13	3,2042	0,81727	0,50504	1,6345	0,71205
14	3,1954	0,81272	0,50432	1,6254	0,71009
15	3,1883	0,80909	0,50375	1,6182	0,70852
16	3,1826	0,80614	0,50329	1,6123	0,70724
17	3,1779	0,80372	0,5029	1,6074	0,70619
18	3,1739	0,80169	0,50258	1,6034	0,70531
19	3,1705	0,79999	0,50232	1,6	0,70456
20	3,1677	0,79854	0,50209	1,5971	0,70393
30	3,1531	0,79118	0,50092	1,5824	0,70069
40	3,1481	0,78864	0,50052	1,5773	0,69957
50	3,1457	0,78747	0,50033	1,5749	0,69905
60	3,1445	0,78684	0,50023	1,5737	0,69877
70	3,1437	0,78645	0,50017	1,5729	0,6986
80	3,1432	0,78621	0,50013	1,5724	0,69849
90	3,1429	0,78604	0,5001	1,5721	0,69842
100	3,1426	0,78592	0,50008	1,5718	0,69836
200	3,1419	0,78553	0,50002	1,5711	0,69819
300	3,1417	0,78546	0,50001	1,5709	0,69816
400	3,1417	0,78543	0,50001	1,5709	0,69815
500	3,1416	0,78542	0,5	1,5708	0,69814
32	3,1517	0,79048	0,50081	1,581	0,70038
64	3,1441	0,78666	0,5002	1,5733	0,69869
128	3,1422	0,78571	0,50005	1,5714	0,69827
256	3,1418	0,78548	0,50001	1,571	0,69817
512	3,1416	0,78542	0,5	1,5708	0,69814
1000	3,1416	0,7854	0,5	1,5708	0,69813

Mechanical Symmetry

Regular Polygonal Tubes – h_1, h_2

$$Area = Area_k \left(h_2^2 - h_1^2 \right)$$

$$I_k = MoI_k \left(h_2^4 - h_1^4 \right) \quad i_k = RadGir_k \sqrt{h_2^2 + h_1^2}$$

$$I_p = PolMoI_k \left(h_2^4 - h_1^4 \right) \quad I_{kap} = ApMoI_k \frac{\left(h_2 - h_1 \right) \left(h_2^2 + h_1 h_2 + h_1^2 \right)^2}{h_2 + h_1}$$

k	$Area_k$	MoI_k	$RadGir_k$	$PolMoI_k$	$ApMoI_k$
3	5,1962	2,5981	0,70711	5,1962	1,1547
4	4	1,3333	0,57735	2,6667	0,88889
5	3,6327	1,068	0,54221	2,136	0,80727
6	3,4641	0,96225	0,52705	1,9245	0,7698
7	3,371	0,9079	0,51897	1,8158	0,74912
8	3,3137	0,87581	0,5141	1,7516	0,73638
9	3,2757	0,8551	0,51092	1,7102	0,72794
10	3,2492	0,84088	0,50872	1,6818	0,72204
12	3,2154	0,82309	0,50595	1,6462	0,71453
13	3,2042	0,81727	0,50504	1,6345	0,71205
14	3,1954	0,81272	0,50432	1,6254	0,71009
15	3,1883	0,80909	0,50375	1,6182	0,70852
16	3,1826	0,80614	0,50329	1,6123	0,70724
17	3,1779	0,80372	0,5029	1,6074	0,70619
18	3,1739	0,80169	0,50258	1,6034	0,70531
19	3,1705	0,79999	0,50232	1,6	0,70456
20	3,1677	0,79854	0,50209	1,5971	0,70393
30	3,1531	0,79118	0,50092	1,5824	0,70069
40	3,1481	0,78864	0,50052	1,5773	0,69957
50	3,1457	0,78747	0,50033	1,5749	0,69905
60	3,1445	0,78684	0,50023	1,5737	0,69877
70	3,1437	0,78645	0,50017	1,5729	0,6986
80	3,1432	0,78621	0,50013	1,5724	0,69849
90	3,1429	0,78604	0,5001	1,5721	0,69842
100	3,1426	0,78592	0,50008	1,5718	0,69836
200	3,1419	0,78553	0,50002	1,5711	0,69819
300	3,1417	0,78546	0,50001	1,5709	0,69816
400	3,1417	0,78543	0,50001	1,5709	0,69815
500	3,1416	0,78542	0,5	1,5708	0,69814
32	3,1517	0,79048	0,50081	1,581	0,70038
64	3,1441	0,78666	0,5002	1,5733	0,69869
128	3,1422	0,78571	0,50005	1,5714	0,69827
256	3,1418	0,78548	0,50001	1,571	0,69817
512	3,1416	0,78542	0,5	1,5708	0,69814
1000	3,1416	0,7854	0,5	1,5708	0,69813

Regular Polygonal Tubes – R_1, R_2

$$Area = Area_k \left(R_2^2 - R_1^2 \right)$$

$$I_k = MoI_k \left(R_2^4 - R_1^4 \right) \quad i_k = RadGir_k \sqrt{R_2^2 + R_1^2}$$

$$I_p = PolMoI_k \left(R_2^4 - R_1^4 \right) \quad I_{kap} = ApMoI_k \frac{(R_2 - R_1)(R_2^2 + R_1 R_2 + R_1^2)^2}{R_2 + R_1}$$

k	$Area_k$	MoI_k	$RadGir_k$	$PolMoI_k$	$ApMoI_k$
3	1,299	0,16238	0,35355	0,32476	0,072169
4	2	0,33333	0,40825	0,66667	0,22222
5	2,3776	0,4575	0,43865	0,915	0,34582
6	2,5981	0,54127	0,45644	1,0825	0,43301
7	2,7364	0,59825	0,46757	1,1965	0,49361
8	2,8284	0,63807	0,47497	1,2761	0,53649
9	2,8925	0,66674	0,48011	1,3335	0,5676
10	2,9389	0,68796	0,48382	1,3759	0,59073
12	3	0,71651	0,48871	1,433	0,62201
13	3,0207	0,72634	0,49036	1,4527	0,63282
14	3,0372	0,73423	0,49168	1,4685	0,64151
15	3,0505	0,74065	0,49274	1,4813	0,64859
16	3,0615	0,74595	0,49362	1,4919	0,65443
17	3,0706	0,75036	0,49434	1,5007	0,65931
18	3,0782	0,75408	0,49495	1,5082	0,66341
19	3,0846	0,75723	0,49546	1,5145	0,66691
20	3,0902	0,75994	0,4959	1,5199	0,6699
30	3,1187	0,77399	0,49818	1,548	0,68547
40	3,1287	0,77896	0,49897	1,5579	0,69098
50	3,1333	0,78127	0,49934	1,5625	0,69355
60	3,1359	0,78253	0,49954	1,5651	0,69495
70	3,1374	0,78329	0,49966	1,5666	0,69579
80	3,1384	0,78378	0,49974	1,5676	0,69634
90	3,139	0,78412	0,4998	1,5682	0,69672
100	3,1395	0,78437	0,49984	1,5687	0,69698
200	3,1411	0,78514	0,49996	1,5703	0,69784
300	3,1414	0,78528	0,49998	1,5706	0,698
400	3,1415	0,78533	0,49999	1,5707	0,69806
500	3,1415	0,78536	0,49999	1,5707	0,69809
32	3,1214	0,77536	0,4984	1,5507	0,68699
64	3,1365	0,78288	0,4996	1,5658	0,69533
128	3,1403	0,78477	0,4999	1,5695	0,69743
256	3,1413	0,78524	0,49997	1,5705	0,69796
512	3,1415	0,78536	0,49999	1,5707	0,69809
1000	3,1416	0,78539	0,5	1,5708	0,69812

Mechanical Symmetry

Regular Polygonal Tubes – b_1, b_2

$$Area = Area_k \left(b_2^2 - b_1^2\right)$$

$$I_k = MoI_k \left(b_2^4 - b_1^4\right) \quad i_k = RadGir_k \sqrt{b_2^2 + b_1^2}$$

$$I_p = PolMoI_k \left(b_2^4 - b_1^4\right) \quad I_{kap} = ApMoI_k \frac{\left(b_2 - b_1\right)\left(b_2^2 + b_1 b_2 + b_1^2\right)^2}{b_2 + b_1}$$

k	$Area_k$	MoI_k	$RadGir_k$	$PolMoI_k$	$ApMoI_k$
3	0,433	0,018042	0,20412	0,036084	0,008019
4	1	0,083333	0,28868	0,16667	0,055556
5	1,7205	0,23955	0,37314	0,4791	0,18107
6	2,5981	0,54127	0,45644	1,0825	0,43301
7	3,6339	1,055	0,53882	2,1101	0,87051
8	4,8284	1,8595	0,62057	3,719	1,5635
9	6,1818	3,0453	0,70187	6,0906	2,5925
10	7,6942	4,7153	0,78284	9,4307	4,0489
12	11,196	9,9796	0,94411	19,959	8,6635
13	13,186	13,84	1,0245	27,68	12,058
14	15,335	18,717	1,1048	37,433	16,353
15	17,642	24,773	1,185	49,546	21,694
16	20,109	32,184	1,2651	64,369	28,236
17	22,735	41,138	1,3451	82,276	36,146
18	25,521	51,834	1,4251	103,67	45,602
19	28,465	64,483	1,5051	128,97	56,791
20	31,569	79,31	1,585	158,62	69,913
30	71,358	405,21	2,383	810,41	358,86
40	127,06	1284,8	3,1798	2569,5	1139,7
50	198,68	3141,3	3,9763	6282,6	2788,6
60	286,22	6519	4,7725	13038	5789,4
70	389,67	12083	5,5686	24166	10733
80	509,03	20620	6,3646	41240	18319
90	644,32	33036	7,1605	66072	29353
100	795,51	50360	7,9564	1,01E+05	44750
200	3182,8	8,06E+05	15,915	1,61E+06	7,17E+05
300	7161,7	4,08E+06	23,873	8,16E+06	3,63E+06
400	12732	1,29E+07	31,831	2,58E+07	1,15E+07
500	19894	3,15E+07	39,788	6,30E+07	2,80E+07
32	81,225	525,02	2,5424	1050	465,18
64	325,69	8441	5,0909	16882	7497,1
128	1303,5	1,35E+05	10,185	2,70E+05	1,20E+05
256	5214,9	2,16E+06	20,371	4,33E+06	1,92E+06
512	20860	3,46E+07	40,743	6,93E+07	3,08E+07
1000	79577	5,04E+08	79,577	1,01E+09	4,48E+08

Mechanical Symmetry

Regular Polygonal Tubes

Accuracy $\Delta I_k = \dfrac{I_k - I_{kap}}{I_{kap}} = \Delta I_k$ as a function of $\dfrac{t}{R}$

K \ t/R	0,95	0,9	0,8	0,75	0,7	0,6	0,5	0,4	0,3	0,25	0,1	0,01
3	-0,951	-0,791	-0,152	0,25	0,627	1,122	1,25	1,191	1,102	1,066	1,008	1
4	0,205	0,326	0,467	0,494	0,5	0,475	0,432	0,393	0,363	0,353	0,336	0,333
5	0,268	0,302	0,323	0,317	0,305	0,273	0,24	0,214	0,195	0,189	0,178	0,176
6	0,236	0,248	0,244	0,233	0,22	0,19	0,163	0,142	0,127	0,121	0,112	0,111
7	0,208	0,212	0,2	0,188	0,175	0,147	0,123	0,104	0,091	0,086	0,079	0,077
8	0,188	0,189	0,173	0,161	0,148	0,122	0,099	0,082	0,07	0,066	0,058	0,057
9	0,175	0,173	0,156	0,144	0,131	0,106	0,084	0,068	0,056	0,052	0,045	0,044
10	0,165	0,162	0,144	0,132	0,119	0,095	0,074	0,058	0,047	0,043	0,036	0,035
12	0,152	0,147	0,128	0,116	0,104	0,081	0,061	0,046	0,035	0,031	0,025	0,024
13	0,147	0,143	0,123	0,111	0,099	0,076	0,057	0,042	0,031	0,028	0,021	0,02
14	0,144	0,139	0,119	0,108	0,095	0,072	0,053	0,039	0,028	0,025	0,018	0,017
15	0,141	0,136	0,116	0,104	0,092	0,07	0,051	0,036	0,026	0,022	0,016	0,015
16	0,139	0,133	0,114	0,102	0,09	0,067	0,048	0,034	0,024	0,02	0,014	0,013
17	0,137	0,131	0,112	0,1	0,088	0,065	0,047	0,032	0,022	0,019	0,013	0,012
18	0,135	0,13	0,11	0,098	0,086	0,064	0,045	0,031	0,021	0,017	0,011	0,01
19	0,134	0,128	0,108	0,097	0,085	0,062	0,044	0,03	0,02	0,016	0,01	0,009
20	0,133	0,127	0,107	0,095	0,084	0,061	0,043	0,029	0,019	0,015	0,009	0,008
30	0,127	0,121	0,101	0,089	0,077	0,056	0,037	0,024	0,014	0,01	0,005	0,004
40	0,125	0,119	0,099	0,087	0,075	0,054	0,036	0,022	0,012	0,009	0,003	0,002
50	0,124	0,118	0,098	0,086	0,074	0,053	0,035	0,021	0,011	0,008	0,002	0,001
60	0,124	0,117	0,097	0,085	0,074	0,052	0,034	0,021	0,011	0,008	0,002	0,001
70	0,123	0,117	0,097	0,085	0,073	0,052	0,034	0,02	0,011	0,007	0,002	0,001
80	0,123	0,117	0,096	0,085	0,073	0,052	0,034	0,02	0,011	0,007	0,001	0,001
90	0,123	0,116	0,096	0,085	0,073	0,052	0,034	0,02	0,01	0,007	0,001	<0,001
100	0,123	0,116	0,096	0,085	0,073	0,051	0,034	0,02	0,01	0,007	0,001	<0,001
200	0,123	0,116	0,096	0,084	0,073	0,051	0,033	0,02	0,01	0,007	0,001	<0,001
300	0,123	0,116	0,096	0,084	0,073	0,051	0,033	0,02	0,01	0,007	0,001	<0,001
400	0,122	0,116	0,096	0,084	0,073	0,051	0,033	0,02	0,01	0,007	0,001	<0,001
500	0,122	0,116	0,096	0,084	0,073	0,051	0,033	0,02	0,01	0,007	0,001	<0,001
32	0,127	0,12	0,1	0,089	0,077	0,055	0,037	0,023	0,013	0,01	0,004	0,003
64	0,123	0,117	0,097	0,085	0,074	0,052	0,034	0,02	0,011	0,007	0,002	0,001
128	0,123	0,116	0,096	0,084	0,073	0,051	0,033	0,02	0,01	0,007	0,001	<0,001
256	0,123	0,116	0,096	0,084	0,073	0,051	0,033	0,02	0,01	0,007	0,001	<0,001
512	0,122	0,116	0,096	0,084	0,073	0,051	0,033	0,02	0,01	0,007	0,001	<0,001
1000	0,122	0,116	0,096	0,084	0,073	0,051	0,033	0,02	0,01	0,007	0,001	<0,001

Mechanical Symmetry

6.4 Regular Polygonal Stars

Mechanical Symmetry

Area	Regular Polygonal Stars
bhk	$I_{xy}=0 \Rightarrow I_u = I_v = I_x = I_y = I_k$
$2R^2 k \cos\frac{\pi}{k} \sin\frac{\pi}{k}$	$i_{xy}=0 \Rightarrow i_u = i_v = i_x = i_y = i_k$
$\frac{1}{2} kRp^2 \tan\frac{\pi}{k}$	
$\dfrac{b^2 k}{2\tan\frac{\pi}{k}}$	

I_k	
$\left(\frac{7}{12} bh^3 + \frac{1}{48} b^3 h\right) k$	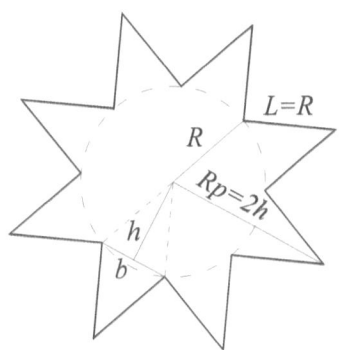
$\frac{1}{6} \cos\frac{\pi}{k} \sin\frac{\pi}{k}\left(\sin^2\frac{\pi}{k} + 7\cos^2\frac{\pi}{k}\right) kR^4$	$L=R$
$\dfrac{\left(1 + 6\cos^2\frac{\pi}{k}\right)\sin\frac{\pi}{k}}{96 \cos^3\frac{\pi}{k}} kR_p^4$	$Rp=2h$
$\dfrac{\left(7\cos\frac{\pi}{k} - 6\cos\frac{\pi}{k}\sin^2\frac{\pi}{k}\right)}{96 \sin^3\frac{\pi}{k}} kb^4$	

i_k	$I_p = I_0$
$\sqrt{\dfrac{28h^2 + b^2}{48}}$	$\dfrac{bh(28h^2 + b^2)k}{24}$
$\sqrt{\dfrac{\sin^2\frac{\pi}{k} + 7\cos^2\frac{\pi}{k}}{12}} R$	$\dfrac{\cos\frac{\pi}{k} \sin\frac{\pi}{k}\left(\sin^2\frac{\pi}{k} + 7\cos^2\frac{\pi}{k}\right) kR^4}{3}$
$\sqrt{\dfrac{\tan^2\frac{\pi}{k} + 7}{48}} R_p$	$\dfrac{\tan\frac{\pi}{k}\left(7 + \tan^2\frac{\pi}{k}\right) kR_p^4}{48}$
$\sqrt{\dfrac{\tan^2\frac{\pi}{k} + 7}{48 \tan\frac{\pi}{k}}} b$	$\dfrac{\left(7 + \tan^2\frac{\pi}{k}\right)}{48 \tan^3\frac{\pi}{k}} kb^4$

Approximate

ΔI_k	I_{kap}
$\dfrac{1}{6} + \dfrac{b^2}{24h^2}$	$0.5 bh^3 k$
	$kR^4 \cos^3\frac{\pi}{k} \sin\frac{\pi}{k}$
$\dfrac{1}{6}\left(1 + \tan^2\frac{\pi}{k}\right)$	$kR_p^4 \frac{1}{16} \tan\frac{\pi}{k}$
	$\dfrac{kb^4}{16 \tan^3\frac{\pi}{k}}$

Mechanical Symmetry

Area	Regular Octagonal Stars
$5.65685R^2$	
$1.65685R_p^2$	$I_{xy}=0 \Rightarrow I_u = I_v = I_x = I_y = I_k$
$9.65685b^2$	$i_{xy}=0 \Rightarrow i_u = i_v = i_x = i_y = i_k$

I_k	
$2.88562R^4$	
$0.2475R_p^4$	$L=R$, R, $R_p=2h$, h, b
$8.40931b^4$	

i_k	$I_p = I_0$
$0.7142R$	$5.77124R^4$
$0.3865R_p$	$0.4951R_p^4$
$0.9332b$	$16.8186b^4$

Approximate

ΔI_k	I_{kap}
0.1953	$2.41421R^4$
	$0.2071R_p^4$
	$7.03553b^4$

Regular Polygonal Stars – R

$$Area = R^2 Area_k \quad I_k = R^4 MoI_k \quad i_k = R \cdot RadGir_k$$
$$I_p = R^4 PolMoI_k \quad I_{kap} = R^4 ApMoI_k$$

k	$Area_k$	MoI_k	$RadGir_k$	$PolMoI_k$	$ApMoI_k$	ApMoI Err
3	2,5981	0,54127	0,45644	1,0825	0,32476	66,7%
4	4	1,3333	0,57735	2,6667	1	33,3%
5	4,7553	1,9525	0,64077	3,9049	1,5562	25,5%
6	5,1962	2,3816	0,677	4,7631	1,9486	22,2%
7	5,4728	2,6773	0,69943	5,3547	2,2213	20,5%
8	5,6569	2,8856	0,71422	5,7712	2,4142	19,5%
9	5,7851	3,0363	0,72446	6,0725	2,5542	18,9%
10	5,8779	3,1481	0,73184	6,2962	2,6583	18,4%
12	6	3,299	0,74151	6,5981	2,799	17,9%
13	6,0414	3,3511	0,74478	6,7023	2,8477	17,7%
14	6,0744	3,393	0,74738	6,786	2,8868	17,5%
15	6,101	3,4271	0,74948	6,8542	2,9187	17,4%
16	6,1229	3,4552	0,7512	6,9104	2,9449	17,3%
17	6,1411	3,4786	0,75263	6,9573	2,9669	17,2%
18	6,1564	3,4984	0,75383	6,9968	2,9854	17,2%
19	6,1693	3,5152	0,75484	7,0304	3,0011	17,1%
20	6,1803	3,5296	0,75571	7,0592	3,0145	17,1%
30	6,2374	3,6044	0,76018	7,2088	3,0846	16,9%
40	6,2574	3,6309	0,76174	7,2618	3,1094	16,8%
50	6,2667	3,6432	0,76247	7,2864	3,121	16,7%
60	6,2717	3,6499	0,76287	7,2998	3,1273	16,7%
70	6,2748	3,654	0,7631	7,3079	3,1311	16,7%
80	6,2767	3,6566	0,76326	7,3132	3,1335	16,7%
90	6,2781	3,6584	0,76336	7,3168	3,1352	16,7%
100	6,2791	3,6597	0,76344	7,3194	3,1364	16,7%
200	6,2822	3,6638	0,76368	7,3276	3,1403	16,7%
300	6,2827	3,6646	0,76373	7,3292	3,141	16,7%
400	6,2829	3,6648	0,76374	7,3297	3,1413	16,7%
500	6,283	3,665	0,76375	7,3299	3,1414	16,7%
32	6,2429	3,6117	0,76061	7,2234	3,0915	16,8%
64	6,2731	3,6518	0,76297	7,3035	3,129	16,7%
128	6,2807	3,6618	0,76357	7,3237	3,1384	16,7%
256	6,2826	3,6644	0,76371	7,3287	3,1408	16,7%
512	6,283	3,665	0,76375	7,33	3,1414	16,7%
1000	6,2831	3,6651	0,76376	7,3303	3,1415	16,7%

Mechanical Symmetry

Regular Polygonal Stars – R_p

$$Area = R_p^2 Area_k \quad I_k = R_p^4 MoI_k$$

$$i_k = R_p RadGir_k \quad I_p = R_p^4 PolMoI_k \quad I_{kap} = R_p^4 ApMoI_k$$

k	$Area_k$	MoI_k	$RadGir_k$	$PolMoI_k$	$ApMoI_k$	ApMoI Err
3	2,5981	0,54127	0,45644	1,0825	0,32476	66,7%
4	2	0,33333	0,40825	0,66667	0,25	33,3%
5	1,8164	0,28486	0,39602	0,56972	0,22704	25,5%
6	1,7321	0,26462	0,39087	0,52924	0,21651	22,2%
7	1,6855	0,25395	0,38816	0,50789	0,21069	20,5%
8	1,6569	0,24755	0,38653	0,49509	0,20711	19,5%
9	1,6379	0,24338	0,38548	0,48675	0,20473	18,9%
10	1,6246	0,24049	0,38475	0,48099	0,20307	18,4%
12	1,6077	0,23686	0,38383	0,47372	0,20096	17,9%
13	1,6021	0,23567	0,38353	0,47134	0,20026	17,7%
14	1,5977	0,23473	0,3833	0,46947	0,19971	17,5%
15	1,5942	0,23398	0,38311	0,46797	0,19927	17,4%
16	1,5913	0,23338	0,38296	0,46675	0,19891	17,3%
17	1,5889	0,23288	0,38283	0,46575	0,19862	17,2%
18	1,5869	0,23246	0,38273	0,46491	0,19837	17,2%
19	1,5853	0,2321	0,38264	0,46421	0,19816	17,1%
20	1,5838	0,23181	0,38256	0,46361	0,19798	17,1%
30	1,5766	0,23028	0,38218	0,46056	0,19707	16,9%
40	1,574	0,22975	0,38205	0,4595	0,19675	16,8%
50	1,5729	0,22951	0,38199	0,45901	0,19661	16,7%
60	1,5722	0,22937	0,38196	0,45875	0,19653	16,7%
70	1,5719	0,22929	0,38194	0,45859	0,19648	16,7%
80	1,5716	0,22924	0,38192	0,45849	0,19645	16,7%
90	1,5714	0,22921	0,38191	0,45841	0,19643	16,7%
100	1,5713	0,22918	0,38191	0,45836	0,19641	16,7%
200	1,5709	0,2291	0,38189	0,4582	0,19637	16,7%
300	1,5709	0,22909	0,38188	0,45817	0,19636	16,7%
400	1,5708	0,22908	0,38188	0,45816	0,19635	16,7%
500	1,5708	0,22908	0,38188	0,45816	0,19635	16,7%
32	1,5759	0,23013	0,38215	0,46026	0,19698	16,8%
64	1,5721	0,22934	0,38195	0,45868	0,19651	16,7%
128	1,5711	0,22914	0,3819	0,45828	0,19639	16,7%
256	1,5709	0,22909	0,38189	0,45818	0,19636	16,7%
512	1,5708	0,22908	0,38188	0,45816	0,19635	16,7%
1000	1,5708	0,22908	0,38188	0,45815	0,19635	16,7%

Mechanical Symmetry

Regular Polygonal Stars – b

$$Area = b^2 Area_k \quad I_k = b^4 MoI_k$$

$$i_k = b \cdot RadGir_k \quad I_p = b^4 PolMoI_k \quad I_{kap} = b^4 ApMoI_k$$

k	Area$_k$	MoI$_k$	RadGir$_k$	PolMoI$_k$	ApMoI$_k$	ApMoI Err
3	0,86603	0,060141	0,26352	0,12028	0,036084	66,7%
4	2	0,33333	0,40825	0,66667	0,25	33,3%
5	3,441	1,0223	0,54507	2,0446	0,81483	25,5%
6	5,1962	2,3816	0,677	4,7631	1,9486	22,2%
7	7,2678	4,7216	0,80601	9,4432	3,9173	20,5%
8	9,6569	8,4093	0,93317	16,819	7,0355	19,5%
9	12,364	13,868	1,0591	27,736	11,666	18,9%
10	15,388	21,577	1,1841	43,155	18,22	18,4%
12	22,392	45,95	1,4325	91,899	38,986	17,9%
13	26,372	63,854	1,5561	127,71	54,261	17,7%
14	30,669	86,493	1,6793	172,99	73,589	17,5%
15	35,285	114,63	1,8024	229,25	97,622	17,4%
16	40,219	149,08	1,9253	298,15	127,06	17,3%
17	45,471	190,71	2,048	381,43	162,66	17,2%
18	51,042	240,47	2,1706	480,95	205,21	17,2%
19	56,93	299,34	2,293	598,68	255,56	17,1%
20	63,138	368,36	2,4154	736,72	314,61	17,1%
30	142,72	1887	3,6362	3774	1614,9	16,9%
40	254,12	5988,5	4,8544	11977	5128,5	16,8%
50	397,36	14648	6,0715	29297	12549	16,7%
60	572,43	30406	7,2882	60812	26052	16,7%
70	779,34	56366	8,5045	1,13E+05	48300	16,7%
80	1018,1	96197	9,7206	1,92E+05	82437	16,7%
90	1288,6	1,54E+05	10,937	3,08E+05	1,32E+05	16,7%
100	1591	2,35E+05	12,153	4,70E+05	2,01E+05	16,7%
200	6365,7	3,76E+06	24,31	7,52E+06	3,22E+06	16,7%
300	14323	1,90E+07	36,466	3,81E+07	1,63E+07	16,7%
400	25464	6,02E+07	48,622	1,20E+08	5,16E+07	16,7%
500	39788	1,47E+08	60,778	2,94E+08	1,26E+08	16,7%
32	162,45	2445,6	3,88	4891,2	2093,3	16,8%
64	651,37	39373	7,7747	78746	33737	16,7%
128	2607,1	6,31E+05	15,557	1,26E+06	5,41E+05	16,7%
256	10430	1,01E+07	31,117	2,02E+07	8,66E+06	16,7%
512	41721	1,62E+08	62,236	3,23E+08	1,39E+08	16,7%
1000	1,59E+05	2,35E+09	121,56	4,70E+09	2,02E+09	16,7%

Mechanical Symmetry

Appendixes

Mechanical Symmetry

Appendix 1: Computer Programs for Chapter 2

Mechanical Symmetry

Mechanical Symmetry

Ap. 1.1 Showing MoI Changing with Rotation and Translation

We have prepared a program called Steiner to illustrate how the MoI changes when moving a section.

Two movements can be done on a section: translation and rotation. You can use the arrow keys to interactively move (up and down to translate, left and right to rotate) and see how MoI value changes.

When translating sections, MoI value changes over "Steiner's parabola". Both sections have the same area, so the parabola is the same, with the only difference of the value at the coordinate origin.

When rotating sections, the MoI value changes over a trigonometric curve for the rectangle and is constant for the circle. Pressing the Z key sets the angle for the section to 0.

Combining shift and control keys with the keys already mentioned, you can change the view, and you can reset the view to original by pressing the R key.

Pressing B makes the window background turn black, and pressing W makes it white.

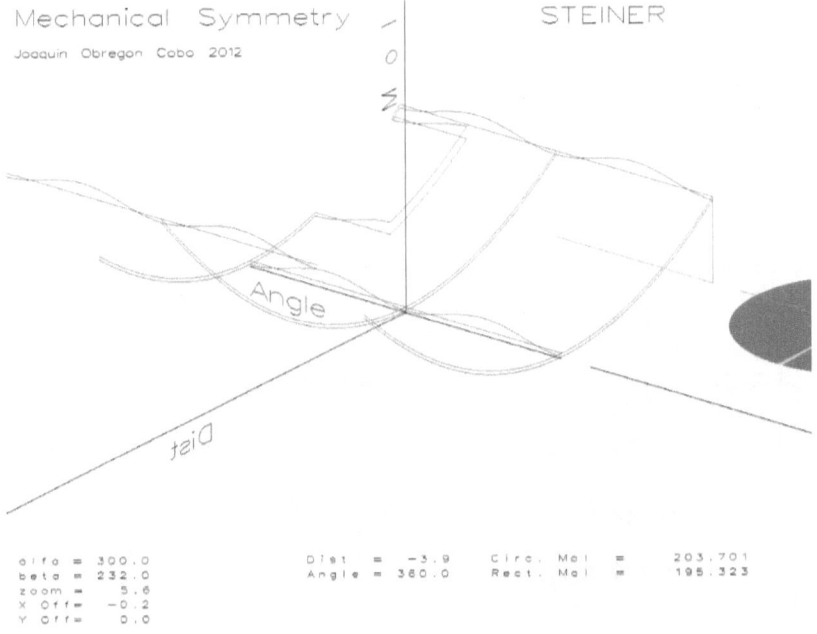

Fig. 35 Program STEINER – MoI Change Visualization (B&W Version)

Mechanical Symmetry

Source Code (in C)

```c
/*****************************************************************

   CopyRight Joaquin Obregon Cobo 2011

   Mechanical Symmetry - Sample program to illustrate MoI Steiner
Theorem  and section rotation

   Disclaimer: Use at you own risk. Author does not accept any
responsability from the use of this code.
*****************************************************************/
// Includes
#include <windows.h>
#include <gl/gl.h>
#include <math.h>

// Defines
#define NUMSECTIONS 2
#define MAXPOINTS 100000
// This "function" to clean a little source code
#define BoFontPrint( fmt , val ) \
    glTranslated( 0 , -1.5 , 0.0 ); \
    sprintf( strinfo , fmt , val ); \
    BoFontString(strinfo);

// Types
typedef struct moi_point {
    float moicir;
    float moirec;
    float dist;
    float angle;
} point;

// Function Declarations
LRESULT CALLBACK WndProc (HWND hWnd, UINT message,WPARAM wParam, LPARAM lParam);
void EnableOpenGL (HWND hWnd, HDC *hDC, HGLRC *hRC);
void DisableOpenGL (HWND hWnd, HDC hDC, HGLRC hRC);
// Stroke Font handling
void BoFontString(const char str[]);
void BoFontFixStep(void);
void BoFontVarStep(void);
void BoFontInit(void);

// Globals
// To animate~change perspective
float alfa = 300.0;
float beta = 220.0;
float zoom =    2.4;
float xOff =    0.0;
float yOff =    0.0;
// To change section's position and orientation
```

Mechanical Symmetry

```c
float dist = 0.0f;
float angle = M_PI;
// To define section's size
const float r = 2.0f;
// To define view
float maxMoI, maxDist, maxAngle;
float black = 0.0f;
float white = 1.0f;

// Array implemented list to hold the points as they are
calculated
// We encapsulate all stack code in "push" and "destroy"
// To store moi - dist - angle
long npoints = 0;
point points[MAXPOINTS];
// We create a new point and add to the list
void *push ( float mc, float mr, float d, float a) {
    if (npoints < MAXPOINTS)  {
       points[npoints].moicir = mc;
       points[npoints].moirec = mr;
       points[npoints].dist = d;
       points[npoints].angle = a;
       npoints++;
    }
    return;
}
// Sequentially go to the end drawing a line with points
void drawLineStrip (){
   long i;
   glBegin(GL_LINE_STRIP);
   glColor3f (0.8f, 0.1f , 0.1f );
   for ( i = 0 ; i < npoints ; i++ ) {
       glVertex3f ( points[i].dist ,
             points[i].angle - M_PI ,
             points[i].moicir/100 );
   }
   glEnd();
   glBegin(GL_LINE_STRIP);
   glColor3f (0.1f, 0.1f , 0.8f );
   for ( i = 0 ; i < npoints ; i++ ) {
       glVertex3f ( points[i].dist ,
             points[i].angle - M_PI ,
             points[i].moirec/100 );
   }
   glEnd();
   // And a cursor
   float maxmoi =  points[npoints-1].moicir > points[npoints-1].moirec ?
           points[npoints-1].moicir :  points[npoints-1].moirec;
   glBegin(GL_LINES);
   glColor3f (0.6f, 0.6f , 0.6f );
```

```
      glVertex3f (dist, 0.0f , 0.0f);
      glVertex3f (dist, angle - M_PI, 0.0f);
      glVertex3f (dist, 3.7f, 0.0f);
      glVertex3f (dist, 18.0f, 0.0f);
      glVertex3f (dist, angle - M_PI, 0.0f);
      glVertex3f (dist, angle - M_PI, maxmoi / 100);
      glEnd();
}
// Sequentially go to the end finding maximums
void findMax (){
   maxMoI = 1.0f;
   maxDist = 1.0f;
   maxAngle = 18; // To allow sections display
   long i;
   for ( i = 0 ; i < npoints ; i++ ) {
       if (points[i].dist > maxDist)
         maxDist = points[i].dist;
       if (points[i].moicir > maxMoI)
         maxMoI = points[i].moicir;
       if (points[i].moirec > maxMoI)
         maxMoI = points[i].moirec;
   }
}
// End of array implemented list

//   Draws sections as needed
void drawSections(void)
{
  // Draw sections in dist-angle plane
  // coordinates are ( dist , angle , moi )
  float rads, inc, y, x;
  // Circle with center at (x,y)
  y = 6.0f;
  x = 0.0f + dist;
  inc = M_PI / 64;
  glBegin (GL_TRIANGLES);
  glPushMatrix();
  glTranslatef( x , y , 0.0f);
  glRotatef( angle * 180.0 / M_PI , 0.0 , 0.0 , 1.0 );
  glColor3f (0.8f, 0.8f , 0.8f );// To show rotation
  for ( rads = angle ; rads < 2.0 * M_PI + angle ;  ) {
    glVertex3f ( x , y , 0.0f );
    glVertex3f ( x + r*cos(rads),y+r*sin(rads),0.0f);
    rads += inc;
    glVertex3f ( x + r*cos(rads),y+r*sin(rads),0.0f);
    glColor3f (0.8f, 0.0f , 0.0f );
  }
  glEnd ();
  glPopMatrix();

  // Rectangle with the same area as circle and sides: l=r
, L=pi*r
```

```c
    y = 14.0f;
    glColor3f (0.1f, 0.1f , 0.8f );
    glPushMatrix();
    glTranslatef( x , y , 0.0f);
    glRotatef( angle * 180.0 / M_PI , 0.0 , 0.0 , 1.0 );
    glBegin (GL_POLYGON);
    glVertex3f ( -r/2.0 , -r * M_PI / 2.0 , 0.0f );
    glVertex3f (  r/2.0 , -r * M_PI / 2.0 , 0.0f );
    glVertex3f (  r/2.0 ,  r * M_PI / 2.0 , 0.0f );
    glVertex3f ( -r/2.0 ,  r * M_PI / 2.0 , 0.0f );
    glEnd ();
    glPopMatrix();
}

// Calculations for the sections
float circle_area (void) {
    return r * r * M_PI;
}
// If both sections have the same area Steiners parabola
is the same
float rect_area (void) {
    return M_PI * r * r;
}
float circle_moi (void) {
    float moi = M_PI * r * r * r * r / 4.0f;
    float steiner = circle_area() * dist * dist;
    return moi + steiner;
}
float rect_moi (void) {
    // Principal moments bh^3 / 12
    float moi1 = r * M_PI * r * r * r / 12.0f;
    float moi2 = r * r * M_PI * r * M_PI * r * M_PI / 12.0f;
    // Moment with the rotation
    float moi = moi1 * cos(angle)*cos(angle) + moi2 * sin(angle)*sin(angle);
    float steiner = rect_area() * dist * dist;
    return moi + steiner;
}

// WinMain
int WINAPI WinMain (HINSTANCE hInstance,
        HINSTANCE hPrevInstance,
        LPSTR lpCmdLine,
        int iCmdShow)
{
  WNDCLASS wc;
  HWND hWnd;
  HDC hDC;
  HGLRC hRC;
  MSG msg;
  BOOL bQuit = FALSE;
```

```
float cubo = 0.0;
char strinfo[32];// We use it with BoFontPrint macro

/* register window class */
wc.style = CS_OWNDC;
wc.lpfnWndProc = WndProc;
wc.cbClsExtra = 0;
wc.cbWndExtra = 0;
wc.hInstance = hInstance;
wc.hIcon = LoadIcon (NULL, IDI_APPLICATION);
wc.hCursor = LoadCursor (NULL, IDC_ARROW);
wc.hbrBackground=(HBRUSH)GetStockObject (BLACK_BRUSH);
wc.lpszMenuName = NULL;
wc.lpszClassName = "steiner";
RegisterClass (&wc);

/* create main window */
hWnd = CreateWindow (
  "Steiner", "Mechanical Symmetry",
  WS_CAPTION | WS_POPUPWINDOW | WS_VISIBLE,
  0, 0, 1024, 1024 ,
  NULL, NULL, hInstance, NULL);

/* enable OpenGL for the window */
EnableOpenGL (hWnd, &hDC, &hRC);
// Init Font
BoFontInit();

// Init arrays
push( circle_moi(), rect_moi(), dist, angle );

/* program main loop */
while (!bQuit)
{
  /* check for messages */
  if (PeekMessage (&msg, NULL, 0, 0, PM_REMOVE))
  {
    /* handle or dispatch messages */
    if (msg.message == WM_QUIT)
    {
      bQuit = TRUE;
    }
    else
    {
      TranslateMessage (&msg);
      DispatchMessage (&msg);
    }
    // Something happened ...
    // Prepare window
    // Clear
    glClearColor (black, black, black, 1.0f);
    glClear (GL_COLOR_BUFFER_BIT | GL_DEPTH_BUFFER_BIT);
```

Mechanical Symmetry

```
// Draw content
// Set up view
// Set up perspective~projection
glMatrixMode(GL_PROJECTION);
glLoadIdentity();

// Calculate view parameters
findMax();
cubo = maxDist;
if ( cubo < maxAngle ) cubo = maxAngle;
if ( cubo < maxMoI/100 ) cubo = maxMoI/100;
cubo *= 2.0;
glOrtho( -cubo , cubo , -cubo , cubo , -cubo*4.0 , cubo*4.0 );

// Adjust view to users/default parameters
glTranslatef( xOff , yOff , 0.0f );
glRotatef (alfa, 1.0f, 0.0f, 0.0f);
glRotatef (beta, 0.0f, 0.0f, 1.0f);
glScalef ( zoom , zoom , zoom );
float distorsion = 1024 / 728;

// Draw coordinate axes
glColor3f (white, white, white);
glBegin (GL_LINES);
glVertex3f (0.0f, 0.0f, 0.0f); glVertex3f (cubo, 0.0f, 0.0f);
glVertex3f (0.0f, -M_PI, 0.0f); glVertex3f (0.0f, M_PI, 0.0f);
glVertex3f (0.0f, 3.7f, 0.0f); glVertex3f (0.0f, 18.0f, 0.0f);
glVertex3f (0.0f, 0.0f, 0.0f); glVertex3f (0.0f, 0.0f, cubo );
glEnd ();

// Draw axes labels
// Dsitance label
BoFontVarStep();
    float scale = 0.6f;
        glPushMatrix();
    glRotatef( 90, 1.0f, 0.0f , 0.0f);
    glTranslatef( 5.0f , 0.2f , 0.0f );
        glScalef( scale, scale*distorsion , 1.0f);
        BoFontPrint( "Dist", dist );
    glPopMatrix();
    // Angle label
        glPushMatrix();
    glRotatef( 90, 1.0f, 0.0f , 0.0f);
    glRotatef( 90, 0.0f, 1.0f , 0.0f);
    glTranslatef( -3.14f, 0.2f, 0.0f );
        glScalef( scale, scale*distorsion , 1.0f);
```

```
            BoFontPrint( "Angle", angle/M_PI*180);
        glPopMatrix();
        // MoI label
            glPushMatrix();
            glRotatef( 90, 0.0f, 1.0f , 0.0f);
            glRotatef( 90, 1.0f, 0.0f , 0.0f);
            glTranslatef( -5.0f , 0.2f , 0.0f );
                glScalef( -scale*distorsion , scale , 1.0f);
            BoFontString("M o I");
            glPopMatrix();

            drawSections();
            drawLineStrip();

    // Draw Info
    // Display information about view angles
    // Setup View
    glClear ( GL_DEPTH_BUFFER_BIT);
    glMatrixMode(GL_PROJECTION);
    glLoadIdentity();
    // We define a view with row and column characters
coordinates
    glOrtho(0.0 , 70.0 , 0.0 , 64.0, -0.01 , 0.01 );

    BoFontVarStep();
    glColor3f (white, white, white);
    glPushMatrix ();
      glTranslatef( 1.0 , 61.5 , 0.0 );
          glScalef( 2.5f , 2.5f , 1.0f);
        BoFontString( "Mechanical Symmetry     STEINER"
);
      glTranslatef( 0.0 , -1.3 , 0.0 );
          glScalef( 0.5f , 0.5f , 1.0f);
          BoFontString( "Joaquin Obregon Cobo 2012" );
    glPopMatrix ();
    // Display info about view
    BoFontFixStep();
    glPushMatrix ();
      glTranslatef( 1.0 , 8.0 , 0.0 );
      BoFontPrint( "alfa = %5.1f" , alfa );
      BoFontPrint( "beta = %5.1f" , beta );
      BoFontPrint( "zoom = %5.1f" , zoom );
      BoFontPrint( "X Off= %5.1f" , xOff );
      BoFontPrint( "Y Off= %5.1f" , yOff );
    glPopMatrix ();
    // Display DATA
    glPushMatrix ();
      glTranslatef( 26.0 , 8.0 , 0.0 );
      BoFontPrint( "Dist  = %5.1f" , dist );
      BoFontPrint( "Angle = %5.1f" , angle * 180.0f /
M_PI );
    glPopMatrix ();
```

```
      glTranslatef( 42.0 , 8.0 , 0.0 );
      BoFontPrint( "Circ. MoI   = %10.3f" , circle_moi() );
      BoFontPrint( "Rect. MoI   = %10.3f" , rect_moi() );

      glFlush();

      SwapBuffers(hDC);
    }
    else
    {
      // nothing to do
    }
  }

  /* shutdown OpenGL */
  DisableOpenGL (hWnd, hDC, hRC);

  /* destroy the window explicitly */
  DestroyWindow (hWnd);

  return msg.wParam;
}

/*******************
 * Window Procedure
 *
 *******************/
LRESULT CALLBACK WndProc (HWND hWnd, UINT message,
             WPARAM wParam, LPARAM lParam)
{
  static int shifted = 0;
  static int ctrled = 0;
  switch (message)
  {
  case WM_CREATE:
    return 0;
  case WM_CLOSE:
    PostQuitMessage (0);
    return 0;

  case WM_DESTROY:
    return 0;

  case WM_KEYUP:
    switch (wParam)
    {
    case VK_SHIFT:
      shifted = 0;
      break;
    case VK_CONTROL:
      ctrled = 0;
      break;
```

```
      }
      return 0;

  case WM_KEYDOWN:
    switch (wParam)
    {
    case VK_ESCAPE:
      PostQuitMessage(0);
      break;
    case VK_SHIFT:
      shifted = 1;
      break;
    case VK_CONTROL:
      ctrled = 1;
      break;
    case 'B':
      black = 0.0f;
      white = 1.0f;
      break;
    case 'W':
      black = 1.0f;
      white = 0.0f;
      break;
    case 'R':
      alfa = 300.0;
      beta = 220.0;
      zoom =   2.4;
      xOff =   0.0;
      yOff =   0.0;
      break;
    }
    if (shifted) {
      switch (wParam)
      {
      case 'Z':
          zoom += 0.1;
        break;
      case VK_UP:
        alfa += 2;
        alfa = alfa == 360 ? 0 : alfa;
        break;
      case VK_DOWN:
        alfa -= 2;
        alfa = alfa == -2 ? 358 : alfa;
        break;
      case VK_LEFT:
        beta += 2;
        beta = beta == 360 ? 0 : beta;
        break;
      case VK_RIGHT:
        beta -= 2;
        beta = beta == -2 ? 358 : beta;
```

```
      break;
    }
  } else if (ctrled) {
    switch (wParam)
    {
    case 'Z':
        zoom -= 0.1;
        zoom = zoom < 0.1 ? 0.1 : zoom;
      break;
    case VK_UP:
      yOff += 0.1;
      break;
    case VK_DOWN:
      yOff -= 0.1;
      break;
    case VK_LEFT:
      xOff -= 0.1;
      break;
    case VK_RIGHT:
      xOff += 0.1;
      break;
    }
  } else {
    float inc;
    switch (wParam)
    {
    case 'Z':
      for ( inc = (M_PI - angle)/50.0 ;
            fabs(angle - M_PI) > fabs(inc) ;
            angle += inc ) {
        push( circle_moi(), rect_moi(), dist, angle );
        drawLineStrip();
      }
      angle = M_PI;
      push( circle_moi(), rect_moi(), dist, angle );
      break;
    case VK_DOWN:
      dist += 0.1;
      // Add points to graph
      push( circle_moi(), rect_moi(), dist, angle );
      break;
    case VK_UP:
      dist -= 0.1;
      // Add points to graph
      push( circle_moi(), rect_moi(), dist, angle );
      break;
    case VK_RIGHT:
      angle += 0.031416;
      angle = angle >= 2 * M_PI ? 2*M_PI : angle;
      // Add points to graph
      push( circle_moi(), rect_moi(), dist, angle );
      break;
```

```
        case VK_LEFT:
           angle -= 0.031416;
           angle = angle <= 0 ? 0 : angle;
           // Add points to graph
           push( circle_moi(), rect_moi(), dist, angle );
          break;
        }
     }
     return 0;

  default:
     return DefWindowProc (hWnd, message, wParam, lParam);
  }
}
// Enable OpenGL
void EnableOpenGL (HWND hWnd, HDC *hDC, HGLRC *hRC)
{
  PIXELFORMATDESCRIPTOR pfd;
  int iFormat;

  /* get the device context (DC) */
  *hDC = GetDC (hWnd);

  /* set the pixel format for the DC */
  ZeroMemory (&pfd, sizeof (pfd));
  pfd.nSize = sizeof (pfd);
  pfd.nVersion = 1;
  pfd.dwFlags = PFD_DRAW_TO_WINDOW |
     PFD_SUPPORT_OPENGL | PFD_DOUBLEBUFFER;
  pfd.iPixelType = PFD_TYPE_RGBA;
  pfd.cColorBits = 24;
  pfd.cDepthBits = 16;
  pfd.iLayerType = PFD_MAIN_PLANE;
  iFormat = ChoosePixelFormat (*hDC, &pfd);
  SetPixelFormat (*hDC, iFormat, &pfd);

  /* create and enable the render context (RC) */
  *hRC = wglCreateContext( *hDC );
  wglMakeCurrent( *hDC, *hRC );
  // Zbuffering
  glEnable(GL_DEPTH_TEST);

}


// Disable OpenGL
void DisableOpenGL (HWND hWnd, HDC hDC, HGLRC hRC)
{
  wglMakeCurrent (NULL, NULL);
  wglDeleteContext (hRC);
  ReleaseDC (hWnd, hDC);
}
```

Appendix 2: Computer Programs for Chapter 3

Mechanical Symmetry

Ap. 2.2 Calculation of sine² Sums

Basic

```
REM
*************************************************************
REM    Copyright Joaquin Obregon Cobo 2011
REM    Sample program showing constant sums of sen²…
REM
*************************************************************
REM Constants
LET SAMPLES = 16
LET ANGININC = 15.0
LET ANGININCRAD = 0.261799387799149436538553615273299
REM Table Header
PRIN  "------------------------- Mechanical Symmetry ----
----------------"
PRINT "                     Sum(sin2)/k"
PRINT "    /------------------------- Angle -------------
--------------\"
PRINT "  k  ";
FOR angIni = 0.0 TO 90.0 STEP ANGININC
   PRINT USING "######## ": angIni;
NEXT angIni
PRINT
REM Loop FOR k from 1 to 16
REM   Loop FOR orientation from 0 to 90 grados
REM     Loop FOR sum for each particle (i from 1 to k)
FOR k=1.0 TO SAMPLES STEP 1.0
   PRINT USING " ##  ": k;
   REM ang Angle between particles
   LET ang = PI * 2.0 / k
   REM angIni Defines section rotation as the angle for
   REM   the first particle
   FOR angIni = 0.0 TO PI/2.0 STEP ANGININCRAD
    LET sum = 0.0
    FOR i=1 TO k
     REM alfa is the angle for every particle
     LET alfa = ang * i + angIni
     LET sum = sum + SIN(alfa)*SIN(alfa)
    NEXT i
    REM Now sum contains the summation of sin²
    LET sum = sum / k
    PRINT USING "----%.###": sum;
   NEXT angIni
   PRINT
NEXT k
END
```

Output Listing:

```
---------------------- Mechanical Symmetry --------------------
                    Sum(sin2)/k
      /------------------------ Angle ------------------------\
   k     0    15     30     45    60     75     90
   1   0.000 0.067 0.250 0.500 0.750 0.933 1.000
   2   0.000 0.067 0.250 0.500 0.750 0.933 1.000
   3   0.500 0.500 0.500 0.500 0.500 0.500 0.500
   4   0.500 0.500 0.500 0.500 0.500 0.500 0.500
   5   0.500 0.500 0.500 0.500 0.500 0.500 0.500
   6   0.500 0.500 0.500 0.500 0.500 0.500 0.500
   7   0.500 0.500 0.500 0.500 0.500 0.500 0.500
   8   0.500 0.500 0.500 0.500 0.500 0.500 0.500
   9   0.500 0.500 0.500 0.500 0.500 0.500 0.500
  10   0.500 0.500 0.500 0.500 0.500 0.500 0.500
  11   0.500 0.500 0.500 0.500 0.500 0.500 0.500
  12   0.500 0.500 0.500 0.500 0.500 0.500 0.500
  13   0.500 0.500 0.500 0.500 0.500 0.500 0.500
  14   0.500 0.500 0.500 0.500 0.500 0.500 0.500
  15   0.500 0.500 0.500 0.500 0.500 0.500 0.500
  16   0.500 0.500 0.500 0.500 0.500 0.500 0.500
```

C

```c
/*************************************************************

    CopyRight Joaquin Obregon Cobo 2011

    Mechanical Symmetry

    Sample program to illustrate constant sum of sen2

    Disclaimer: Use at you own risk. Author does not accept
    any responsability from the use of this code.

*************************************************************/

#include <stdio.h>
#include <stdlib.h>
#include <math.h>

// Defines
#define SAMPLES 16
#define ANGININC 15.0
#define ANGININCRAD 0.26179938779914943653855361527329

int main(int argc, char *argv[])
{
   float k, angIni, alfa, ang, sum;
   // Table Header
```

```c
  printf("\n-------------------------- Mechanical Symmetry --------------------\n");
  printf("                    Sum(sin2)/k\n\n");
  printf("\n   /-------------------------- Angle ----------------------------\\");
  printf("\n k  ");
  for ( angIni = 0.0f ; angIni <= 90.0f ; angIni += ANGININC ) {
     printf("%9.3f", angIni );
  }
  // Loop for k from 1 to 16
  //    loop for orientation fro 0 to 90 degrees
  //       loop to sum every particle (i from 1 to k)
  for ( k=1.0f ; k<=SAMPLES ; k += 1.0f ) {
    printf("\n%3d  ", (int)k );
    float angIni, alfa, ang, sum;
    // ang is the angle between particles
    ang = M_PI * 2.0f / k;
    // angIni defines the rotation of the section as the initial
    //    angle for the first particle
    for ( angIni = 0.0f ; angIni <= M_PI/2.0 ; angIni += ANGININCRAD ) {
       sum = 0.0f;
       int i;
       for ( i=1 ; i<=k ; i++) {
         // alfa is the angle for every particle
         alfa = ang * i + angIni;
         sum += sin(alfa)*sin(alfa);
       }
       // Now sum has the sum of sin2
       sum /= k;
       // Now it contains the constant sum/k (for k≥ 3)
       // And we print it
       printf("%9.3f", sum );
    }
  }

  // Uncomment to pause at finish
  //printf("\n");
  //system("PAUSE");
  return 0;
}
```

Output Listing

```
---------------------- Mechanical Symmetry --------------------
                    Sum(sin2)/k


      /------------------------- Angle ------------------------\
   k   0.000   15.000   30.000   45.000   60.000   75.000   90.000
   1   0.000   0.067    0.250    0.500    0.750    0.933    1.000
   2   0.000   0.067    0.250    0.500    0.750    0.933    1.000
   3   0.500   0.500    0.500    0.500    0.500    0.500    0.500
   4   0.500   0.500    0.500    0.500    0.500    0.500    0.500
   5   0.500   0.500    0.500    0.500    0.500    0.500    0.500
   6   0.500   0.500    0.500    0.500    0.500    0.500    0.500
   7   0.500   0.500    0.500    0.500    0.500    0.500    0.500
   8   0.500   0.500    0.500    0.500    0.500    0.500    0.500
   9   0.500   0.500    0.500    0.500    0.500    0.500    0.500
  10   0.500   0.500    0.500    0.500    0.500    0.500    0.500
  11   0.500   0.500    0.500    0.500    0.500    0.500    0.500
  12   0.500   0.500    0.500    0.500    0.500    0.500    0.500
  13   0.500   0.500    0.500    0.500    0.500    0.500    0.500
  14   0.500   0.500    0.500    0.500    0.500    0.500    0.500
  15   0.500   0.500    0.500    0.500    0.500    0.500    0.500
  16   0.500   0.500    0.500    0.500    0.500    0.500    0.500
```

If $k \geq 3$ then $\dfrac{\sum_{n=1}^{k} \sin^2\left(\dfrac{2\pi}{k} n + \alpha_0\right)}{k}$ is $\dfrac{1}{2}$

Ap. 2.2 Drawing the sine² Sums

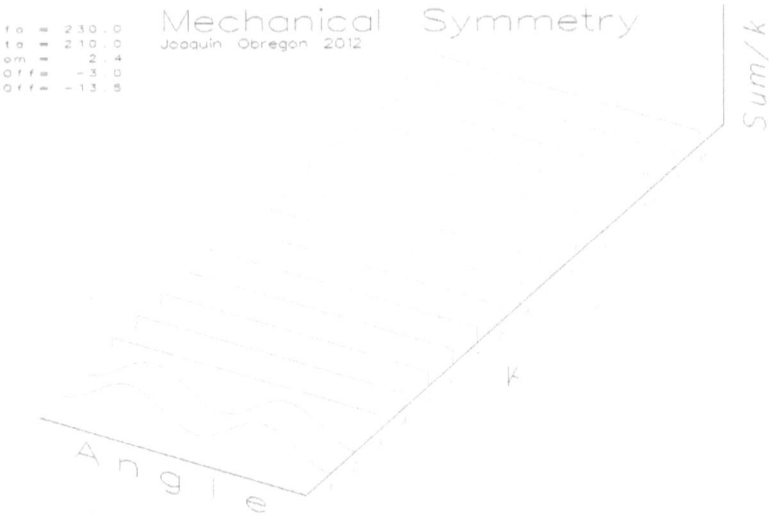

Fig. 36 Sin² Sums (Black and White Version from Color Original)

You can interactively modify the view. Use the arrow keys to rotate and Z to zoom (you can use shift key, too).

```
/************************************************************
    CopyRight Joaquin Obregon Cobo 2011
    Mechanical Symmetry
    Sample program to illustrate constant sum of sen2
    Disclaimer: Use at you own risk. Author does not accept
    any responsability from the use of this code.
************************************************************/

// Includes
#include <windows.h>
#include <gl/gl.h>
#include <gl/glu.h>
#include <math.h>

// Defines
#define SAMPLES 16
#define SAMPLESf 16.0f

// Function Declarations
LRESULT CALLBACK WndProc (HWND hWnd, UINT message,WPARAM
wParam, LPARAM lParam);
```

```c
void EnableOpenGL (HWND hWnd, HDC *hDC, HGLRC *hRC);
void DisableOpenGL (HWND hWnd, HDC hDC, HGLRC hRC);

// Globals
float alfa = 230.0; // To animate~change perspective
float beta = 210.0; // To animate~change perspective
float zoom =   2.4; // To animate~change perspective
float xOff =  -3.0; // To animate~change perspective
float yOff = -13.5; // To animate~change perspective

// WinMain
int WINAPI WinMain (HINSTANCE hInstance, HINSTANCE hPrevInstance,
          LPSTR lpCmdLine, int iCmdShow)
{
  WNDCLASS wc;
  HWND hWnd;
  HDC hDC;
  HGLRC hRC;
  MSG msg;
  BOOL bQuit = FALSE;

  /* register window class */
  wc.style = CS_OWNDC;
  wc.lpfnWndProc = WndProc;
  wc.cbClsExtra = 0;
  wc.cbWndExtra = 0;
  wc.hInstance = hInstance;
  wc.hIcon = LoadIcon (NULL, IDI_APPLICATION);
  wc.hCursor = LoadCursor (NULL, IDC_ARROW);
  wc.hbrBackground = (HBRUSH) GetStockObject (BLACK_BRUSH);
  wc.lpszMenuName = NULL;
  wc.lpszClassName = "Sin2 Sums";
  RegisterClass (&wc);

  /* create main window */
  hWnd = CreateWindow (
    "Sin2 Sums", "Mechanical Symmetry",
    WS_CAPTION | WS_POPUPWINDOW | WS_VISIBLE,
  // Change for a different value to get window size bigger (or smaller)
    0, 0, 1024 , 1024,
    NULL, NULL, hInstance, NULL);

  /* enable OpenGL for the window */
  EnableOpenGL (hWnd, &hDC, &hRC);

  /* program main loop */
  while (!bQuit)
  {
    /* check for messages */
```

```c
if (PeekMessage (&msg, NULL, 0, 0, PM_REMOVE))
{
  /* handle or dispatch messages */
  if (msg.message == WM_QUIT)
  {
    bQuit = TRUE;
  }
  else
  {
    TranslateMessage (&msg);
    DispatchMessage (&msg);
  }
  // Something Happened ...
  // Prepare window
  // Clear
  glClearColor (1.0f, 1.0f, 1.0f, 1.0f);
  glClear (GL_COLOR_BUFFER_BIT);
  glColor3f (0.0f, 0.0f, 0.0f);

  // Set up perspective~projection
  glMatrixMode(GL_PROJECTION);
  glLoadIdentity();
  glOrtho(-20.0,20.0,-17.0,17.0,-170.0,170.0);
  glClearColor (1.0f, 0.7f , 0.7f , 1.0f);

  // Draw content
  BoFontVarStep();
  glPushMatrix();
  glScalef( 2.0,2.0,1.0);
      glTranslated( -10.0 / 2.0 , 15.0 / 2.0 , 0.0 );
  BoFontString("Mechanical Symmetry");
      glTranslated( 0 , -0.5 , 0.0 );
  glScalef( 0.4,0.4,1.0);
  BoFontString("Joaquin Obregon 2012");
  glPopMatrix();

  BoFontFixStep();
      char strinfo[32];
  // Display information about view angles
  glPushMatrix();
  glScalef( 0.6,0.6,0.6);
      glTranslated( -19.0 / 0.6 , 15.0 / 0.6 , 0.0 );
  sprintf( strinfo , "alfa = %5.1f" , alfa );
  BoFontString(strinfo);
      glTranslated( 0 , -1.5 , 0.0 );
  sprintf( strinfo , "beta = %5.1f" , beta );
  BoFontString(strinfo);
      glTranslated( 0 , -1.5 , 0.0 );
  sprintf( strinfo , "zoom = %5.1f" , zoom );
  BoFontString(strinfo);
      glTranslated( 0 , -1.5 , 0.0 );
  sprintf( strinfo , "X Off= %5.1f" , xOff );
```

```
    BoFontString(strinfo);
        glTranslated( 0 , -1.5 , 0.0 );
    sprintf( strinfo , "Y Off= %5.1f" , yOff );
    BoFontString(strinfo);
    glPopMatrix();

    // Set up view
    glPushMatrix ();
    glTranslatef( xOff , yOff , 0.0f );
    glRotatef (alfa, 1.0f, 0.0f, 0.0f);
    glRotatef (beta, 0.0f, 0.0f, 1.0f);
    glScalef ( zoom , zoom , zoom );

    // Draw coordinate axes
    glBegin (GL_LINES);
    glVertex3f (0.0f, 0.0f, 0.0f);  glVertex3f (M_PI*2,
0.0f, 0.0f);
    glVertex3f (0.0f, 0.0f, 0.0f);  glVertex3f (0.0f,
17.0f, 0.0f);
    glVertex3f (0.0f, 17.0f, 0.0f);  glVertex3f (0.0f,
17.0f, 4.0f);
    glEnd ();

    // Draw labels
    // Angle label
        glPushMatrix();
        glScalef( -1.0, 1.0 , 1.0);
      glRotatef( 90, 1.0f, 0.0f , 0.0f);
      glTranslated( - M_PI , -1.0f , 0.0f );
      BoFontCenter();
    BoFontString("Angle");
    BoFontLeft();
        glPopMatrix();
        // Sum/k label
            glPushMatrix();
        glRotatef( 90, 0.0f, 1.0f , 0.0f);
        glRotatef( 90, 1.0f, 0.0f , 0.0f);
        glTranslated( 0.0 , -1.0f , -17.0f );
    glScalef( -0.8 , 0.8 , 0.8 );
        BoFontString("Sum/k");
        glPopMatrix();
    // K label
    glPushMatrix();
    glRotatef( 90, 1.0f, 0.0f , 0.0f);
    glRotatef( 90, 0.0f, 1.0f , 0.0f);
    glTranslated( SAMPLESf/2.0f , -2.4f , 0.0f );
    BoFontString("k");
    glPopMatrix();

    // Here you will find the calculations
    // Loop for k from 1 to 16
    //    loop for orientation fro 0 to 2*PI
```

Mechanical Symmetry

```
//          loop to sum every particle (i from 1 to k)
BoFontCenter();
float k;
for ( k=1.0f ; k<=SAMPLESf ; k += 1.0f ) {
  // Some drawing steps
  glBegin (GL_LINE_STRIP);
  float color = k / SAMPLES / 1.6f;
  glColor3f (1.0f, color , 1.0f - color);
  glVertex3f (0.0f, k , 0.0f);
  float angIni, alfa, ang, sum;
  // ang is the angle between particles
  ang = M_PI * 2.0f / k;
  // angIni defines the rotation of the section as the initial
  //     angle for the first particle
  for ( angIni = 0.0f ; angIni <= 2*M_PI ; angIni += 0.05 ) {
     sum = 0.0f;
     int i;
     for ( i=1 ; i<=k ; i++) {
        // alfa is the angle for every particle
        alfa = ang * i + angIni;
        sum += sin(alfa)*sin(alfa);
     }
     // Now sum has the sum of sin2
     sum /= k;
     // Now it contains the constant sum/k (for k≥ 3)
     // And we draw it
     glVertex3f (angIni, k , sum);
  }
  glVertex3f (M_PI*2.0f, k , 0.0f);
  glEnd ();
  // K index labels
  glPushMatrix();
  glRotatef( 90, 1.0f, 0.0f , 0.0f);
  glRotatef( 90, 0.0f, 1.0f , 0.0f);
  glTranslated( k , -0.6f , 0.0f );
  glScalef( 0.4 , 0.4 , 0.4 );
  char strnum[4];
  sprintf( strnum , "%.0f" , k );
  BoFontString(strnum);
  glPopMatrix();
}
BoFontLeft();
glPopMatrix ();

SwapBuffers (hDC);
  }
  else
  {// Do nothing after nothing happened
  }
}
```

```
  /* shutdown OpenGL */
  DisableOpenGL (hWnd, hDC, hRC);

  /* destroy the window explicitly */
  DestroyWindow (hWnd);

  return msg.wParam;
}


// Window Procedure to handle interaction with user and
system
LRESULT CALLBACK WndProc (HWND hWnd, UINT message,
              WPARAM wParam, LPARAM lParam)
{
  static int shifted = 0;
  switch (message)
  {
  case WM_CREATE:
    return 0;
  case WM_CLOSE:
    PostQuitMessage (0);
    return 0;

  case WM_DESTROY:
    return 0;

  case WM_KEYUP:
    switch (wParam)
    {
    case VK_SHIFT:
      shifted = 0;
      break;
    case 'R':
      alfa = 230.0;
      beta = 210.0;
      zoom =   2.4;
      xOff =  -3.0;
      yOff = -13.5;
      break;
    case VK_ESCAPE:
      PostQuitMessage(0);
      break;
    }
    return 0;

  case WM_KEYDOWN:
    if (shifted) {
      switch (wParam)
      {
      case 'Z':
```

```
        zoom += 0.1;
      break;
    case VK_UP:
      alfa += 2;
      alfa = alfa == 360 ? 0 : alfa;
      break;
    case VK_DOWN:
      alfa -= 2;
      alfa = alfa == -2 ? 358 : alfa;
      break;
    case VK_LEFT:
      beta += 2;
      beta = beta == 360 ? 0 : beta;
      break;
    case VK_RIGHT:
      beta -= 2;
      beta = beta == -2 ? 358 : beta;
      break;
    case VK_SHIFT:
      shifted = 1;
      break;
    }
  } else {
    switch (wParam)
    {
    case 'Z':
      zoom -= 0.1;
      zoom = zoom < 0.1 ? 0.1 : zoom;
      break;
    case VK_UP:
      yOff += 0.25;
      break;
    case VK_DOWN:
      yOff -= 0.25;
      break;
    case VK_LEFT:
      xOff -= 0.25;
      break;
    case VK_RIGHT:
      xOff += 0.25;
      break;
    case VK_SHIFT:
      shifted = 1;
      break;
    }
  }
  return 0;

default:
  return DefWindowProc (hWnd, message, wParam, lParam);
  }
}
```

```c
// Enable OpenGL
void EnableOpenGL (HWND hWnd, HDC *hDC, HGLRC *hRC)
{
  PIXELFORMATDESCRIPTOR pfd;
  int iFormat;

  /* get the device context (DC) */
  *hDC = GetDC (hWnd);

  /* set the pixel format for the DC */
  ZeroMemory (&pfd, sizeof (pfd));
  pfd.nSize = sizeof (pfd);
  pfd.nVersion = 1;
  pfd.dwFlags = PFD_DRAW_TO_WINDOW |
    PFD_SUPPORT_OPENGL | PFD_DOUBLEBUFFER;
  pfd.iPixelType = PFD_TYPE_RGBA;
  pfd.cColorBits = 24;
  pfd.cDepthBits = 16;
  pfd.iLayerType = PFD_MAIN_PLANE;
  iFormat = ChoosePixelFormat (*hDC, &pfd);
  SetPixelFormat (*hDC, iFormat, &pfd);

  /* create and enable the render context (RC) */
  *hRC = wglCreateContext( *hDC );
  wglMakeCurrent( *hDC, *hRC );

}

// Disable OpenGL
void DisableOpenGL (HWND hWnd, HDC hDC, HGLRC hRC)
{
  wglMakeCurrent (NULL, NULL);
  wglDeleteContext (hRC);
  ReleaseDC (hWnd, hDC);
}
```

Ap. 2.3 Particles MoI Comparison

This program shows the difference between MoI, calculated by different methods:

- Exact formula for mechanically symmetric sections
- Approximate formula for MS systems
- Sum of each particles' MoI (Steiner)

Basic

```
REM
***************************************************************
REM    Copyright Joaquin Obregon Cobo 2011
REM    Mechanical Symmetry
REM    Sample PROGRAM TO illustrate constant sum OF sen2
REM    Disclaimer: USE AT you own risk. Author does NOT
REM    accept any responsibility from the USE OF this code.
***************************************************************

REM INPUT ALL DATA
INPUT   PROMPT "Number of particles? " : k
INPUT   PROMPT "Particles radius? " : particleRadius
INPUT   PROMPT "Circle radius? " : circleRadius

REM PRINT PARTICLE'S VALUES
PRINT
LET an = PI *particleRadius*particleRadius
PRINT USING "Area of Particle = ##########.##" : an

LET I_n = an*particleRadius*particleRadius/4
PRINT USING "MoI of Particle  = ##########.##" : I_n

REM PRINT SET OF PARTICLES VALUES
LET IkExact = k * ( I_n + (an * circleRadius *
circleRadius / 2))
PRINT USING "Exact MoI   = ##########.##" : IkExact

LET Ik = k * an * circleRadius * circleRadius / 2
PRINT USING "Approx MoI  = ##########.##" : Ik

REM alfaInc IS THE ANGLE BETWEEN PARTICLES
LET alfaInc = (2.0 / k) * PI
LET IkSum = 0.0
REM SUM ALL THE PARTICLES
FOR i=0 TO k
    REM alfa IS THE ANGLE FOR EACH PARTICLE
    LET alfa = alfaInc * i
    REM dist IS THE DISTANCE FROM PARTICLE TO AXIS
```

```
    LET dist = circleRadius * SIN(alfa)
    REM STEINER
    LET IkSum = IkSum + I_n + an * dist * dist
NEXT i
PRINT USING "MoI from sum   = ##########.##" : IkSum
PRINT

REM CALCULATE DIFFERENCE EXACT - APPROXIMATE
LET diff = IkExact - Ik
LET perthousand = diff / IkExact * 1000.0
PRINT USING "Error Approx   = ##########.##" : diff
PRINT USING "       Deviation = (0/00)---%.##" : perthousand

REM CALCULATE DIFFERENCE EXACT - SUMS
LET diff = IkExact - IkSum
LET perthousand = diff / IkExact * 1000.0
PRINT USING "Error Sum      = ##########.##" : diff
PRINT USING "       Deviation = (0/00)---%.##" : perthousand
END
```

Output Listing

```
Number of particles? 6
Particles radius? 2
Circle radius? 100

Area of Particle =     12.57
MoI of Particle  =     12.57
Exact MoI        =  377066.52
Approx MoI       =  376991.12
MoI from sum     =  377079.08

Error Approx     =      75.40
    Deviation = (0/00)    0.20
Error Sum        =     -12.57
    Deviation = (0/00)   -0.03
```

C

```
/***********************************************************
    CopyRight Joaquin Obregon Cobo 2011
    Mechanical Symmetry
    Sample program to illustrate constant sum of sen2
    Disclaimer: Use at you own risk. Author does not accept
    any responsibility from the use of this code.
***********************************************************/

// Includes
#include <stdio.h>
```

Mechanical Symmetry

```c
#include <stdlib.h>
#include <math.h>
// Action
int main(int argc, char *argv[])
{
  int k;
  float particleRadius, circleRadius;
  // Asking for data
  printf( "Number of particles?\n");
  scanf( "%d" , &k );
  printf ("Particles radius?\n");
  scanf( "%f" , &particleRadius );
  printf ("Circle radius?\n");
  scanf( "%f" , &circleRadius );
  float an, In, Ik, IkExact, IkSum;
  // Calculating Particle's values
  // Area
  an = M_PI *particleRadius*particleRadius;
  printf ("\nArea of Particle = %.3f" , an );
  // Moment of Inertia
  In = an*particleRadius*particleRadius/4;
  printf ("\nMoI of Particle  = %.3f\n \n", In );

  // Calculating Set of particles values
  // Exact MoI with Mechanical Symmetry formula
  IkExact = k * ( In + (an * circleRadius * circleRadius / 2));
  printf ("Exact MoI    = %10.2f\n" , IkExact );
  // Approximate MoI with Mechanical Symmetry formula
  Ik = k * an * circleRadius * circleRadius / 2;
  printf ("Approx. MoI   = %10.2f\n" , Ik );

  // Calculate MoI with sums to compare
  // alfaInc is the angle between particles
  float alfaInc = 2 * M_PI / k;
  IkSum = 0.0;
  int i;
  for ( i=0; i<k ; i++ ) {
    // alfa is the angle for each particle
    float alfa = alfaInc * i;
    // distance from particle to axis
    float dist = circleRadius * sin(alfa);
    // Steiner
    IkSum += In + an * dist * dist;
  }
  printf ("MoI from sum    = %10.2f\n \n", IkSum );
  // Calculate difference Exact-Approximate
  float diff, percent;
  diff = IkExact - Ik;
  percent = diff / IkExact * 1000.0;
  printf ("Error Approx.  = %10.2f \n  Deviation(0/00)= %10.2f\n",diff,percent);
```

```
// Calculate difference Exact-Sums
diff = IkExact - IkSum;
percent = diff / IkExact * 1000.0;
printf ("Error Sum    = %10.2f \n  Deviation(0/00)=
%10.2f\n",diff,percent);
// Comment to eliminate pause at finish
system("PAUSE");
return 0;
}
```

Output Listing

```
Number of particles?
6
Particles radius?
2
Circle radius?
100

Area of Particle = 12.566
MoI of Particle  = 12.566

Exact MoI    =   377066.53
Approx. MoI  =   376991.13
MoI from sum =   377066.56

Error Approx.  =      75.41
  Deviation(0/00)=      0.20
Error Sum    =       -0.03
  Deviation(0/00)=     -0.00
```

Output confirms that the value calculated with the formula and calculated as a sum is the same, confirming the validity of the formulas. Depending on compilers and CPUs used, the numerical error (error sum) may vary.

Ap. 2.4 Particles MoI Comparison – Table

This program shows the difference between MoI, calculated by different methods:
- Exact formula for mechanically symmetric sections
- Approximate formula for MS systems
- Sum of each particles' (Steiner)

The particle's system is a set of sixteen circles, evenly located along a circumference

Particles' diameters get the values 8, 10, 12, 16, 20, 25, 32, 40 and 50—values that are, in millimeters, the usual diameters or steel reinforcing bars (rebars) for concrete. The circle where particles are located gets the following values for its diameter: 20, 25, 40, 50, 60, 70, 80, 90, 100, 120, 140, 150 and 200; in centimeters, those are the usual values for reinforced concrete columns.

C

```c
/***********************************************************
    CopyRight Joaquin Obregon Cobo 2011
    Mechanical Symmetry
***********************************************************/

// Includes
#include <stdio.h>
#include <stdlib.h>
#include <math.h>
// Defines
#define NPR 9   // Number of Particle's Radius
#define NCR 13  // Number of Circle's Radius
// Action
int main(int argc, char *argv[])
{
   // k has no influence on this calculations - we set it to 4
   int k=4;
   // We are working in cm
   float particleRadius[NPR]={0.4,0.5,0.6,0.8,1.0,1.25,1.6,2.0,2.5};
   float circleRadius[NCR]={10,12.5,20,25,30,35,40,45,50,60,70,75,100};
   float an, In, Ik, IkExact, IkSum;
   int p;
   // First, Particle Data
   // Diameter
   printf ("\nParticle  \nDiameter  " );
   for ( p=0 ; p<NPR ; p++ ) {
      printf ("%12.3f" , particleRadius[p]*2.0 );
```

```c
  }
  // Area
  printf ("\nArea    " );
  for ( p=0 ; p<NPR ; p++ ) {
    an = M_PI *particleRadius[p]*particleRadius[p];
    printf ("%12.3f" , an );
  }
  // Moment of Inertia
  printf ("\nMoI     " );
  for ( p=0 ; p<NPR ; p++ ) {
    float pr = particleRadius[p];
    In = M_PI * pr*pr*pr*pr / 4;
    printf ("%12.3f" , In );
  }
  // Then variations on circle radius
  int q;
  for ( q=0 ; q<NCR ; q++ ) {
    // For each Circle Radius we
    // Print Circle Radius
    float cr = circleRadius[q];
    printf("\n\nCircle Diam %6.3f", cr*2.0 );
    // Print Exact MoI for each particle radius
    printf ("\nExact MoI " );
    for ( p=0 ; p<NPR ; p++ ) {
      float pr = particleRadius[p];
      an = M_PI *pr*pr;
      In = an *pr*pr / 4;
      IkExact = k * ( In + (an * cr * cr / 2));
      printf ("%12.3f" , IkExact );
    }
    // Print Approximate MoI for each particle radius
    printf ("\nApprox MoI" );
    for ( p=0 ; p<NPR ; p++ ) {
      float pr = particleRadius[p];
      an = M_PI *pr*pr;
      In = an *pr*pr / 4;
      Ik = k * an * cr * cr / 2;
      printf ("%12.3f" , Ik );
    }
    // Print difference   (ExaxctMoI - SumMoI) for each particle radius
    // Just to confirm there is no difference
    printf ("\nSumMoI Dif" );
    for ( p=0 ; p<NPR ; p++ ) {
      float pr = particleRadius[p];
      an = M_PI *pr*pr;
      In = an *pr*pr / 4;
      IkExact = k * ( In + (an * cr * cr / 2));
      float alfaInc = 2 * M_PI / k;
      IkSum = 0.0;
      int i;
      for ( i=0; i<k ; i++ ) {
```

```c
      float alfa = alfaInc * i;
      float dist = cr * sin(alfa);
      IkSum += In + an * dist * dist;
    }
    printf ("%12.3f" , IkExact-IkSum );
  }
  // Print Exact - Approximate for each particle radius
  // We calculate error as difference.
  // We could have used our formula: inacuraccy = 2 * In
/ (an * cr * cr)
  printf ("\nExact-Appr" );
  for ( p=0 ; p<NPR ; p++ ) {
    float pr = particleRadius[p];
    an = M_PI *pr*pr;
    In = an *pr*pr / 4;
    IkExact = k * ( In + (an * cr * cr / 2));
    Ik = k * an * cr * cr / 2;
    float diff, percent;
    diff = IkExact - Ik;
    printf ("%12.3f" , diff );
  }
  // And we print also the ratio (perthousand)
  printf ("\nDif 0/00  " );
  for ( p=0 ; p<NPR ; p++ ) {
    float pr = particleRadius[p];
    an = M_PI *pr*pr;
    In = an *pr*pr / 4;
    IkExact = k * ( In + (an * cr * cr / 2));
    Ik = k * an * cr * cr / 2;
    float diff, percent;
    diff = IkExact - Ik;
    percent = diff / IkExact * 1000.0;
    printf ("%12.3f" , percent );
  }
 }
 printf("\n");
 return 0;
}
```

Table 8 Output – Particles MoI Comparison

Particle Diameter	0.800	1.000	1.200	1.600	2.000	2.500	3.200
Area	0.503	0.785	1.131	2.011	3.142	4.909	8.042
MoI	0.020	0.049	0.102	0.322	0.785	1.917	5.147
Circle Diam 20.000							
Exact MoI	100.611	157.276	226.602	403.411	631.460	989.418	1629.084
Approx MoI	100.531	157.080	226.195	402.124	628.319	981.748	1608.495
SumMol Dif	0.000	0.000	0.000	0.000	0.000	0.000	0.000
Exact-Appr	0.080	0.196	0.407	1.287	3.142	7.670	20.589
Dif 0/00	0.799	1.248	1.797	3.190	4.975	7.752	12.638
Circle Diam 25.000							
Exact MoI	157.160	245.633	353.836	629.605	984.889	1541.651	2533.863
Approx MoI	157.080	245.437	353.429	628.319	981.748	1533.981	2513.274
SumMol Dif	0.000	0.000	0.000	0.000	0.000	0.000	0.000
Exact-Appr	0.080	0.196	0.407	1.287	3.142	7.670	20.589
Dif 0/00	0.512	0.799	1.151	2.044	3.190	4.975	8.125
Circle Diam 40.000							
Exact MoI	402.204	628.515	905.186	1609.782	2516.416	3934.661	6454.571
Approx MoI	402.124	628.319	904.779	1608.495	2513.274	3926.991	6433.982
SumMol Dif	0.000	0.000	0.000	0.000	0.000	0.000	0.000
Exact-Appr	0.080	0.196	0.407	1.287	3.142	7.670	20.589
Dif 0/00	0.200	0.312	0.450	0.799	1.248	1.949	3.190
Circle Diam 50.000							
Exact MoI	628.399	981.944	1414.124	2514.561	3930.133	6143.593	10073.686
Approx MoI	628.319	981.748	1413.717	2513.274	3926.991	6135.923	10053.097
SumMol Dif	0.000	0.000	0.000	0.000	0.000	0.000	0.000
Exact-Appr	0.080	0.196	0.407	1.287	3.142	7.670	20.589
Dif 0/00	0.128	0.200	0.288	0.512	0.799	1.248	2.044
Circle Diam 60.000							
Exact MoI	904.859	1413.913	2036.159	3620.402	5658.008	8843.399	14497.049
Approx MoI	904.779	1413.717	2035.752	3619.115	5654.867	8835.729	14476.460
SumMol Dif	0.000	0.000	0.000	0.000	0.000	0.000	0.000
Exact-Appr	0.080	0.196	0.407	1.287	3.142	7.670	20.589
Dif 0/00	0.089	0.139	0.200	0.355	0.555	0.867	1.420
Circle Diam 80.000							
Exact MoI	1608.576	2513.470	3619.522	6435.269	10056.238	15715.634	25756.518
Approx MoI	1608.495	2513.274	3619.115	6433.982	10053.097	15707.964	25735.928
SumMol Dif	0.000	0.000	0.000	0.000	0.000	0.000	0.000
Exact-Appr	0.080	0.196	0.407	1.287	3.142	7.670	20.590
Dif 0/00	0.050	0.078	0.113	0.200	0.312	0.488	0.799
Circle Diam 90.000							
Exact MoI	2035.833	3181.059	4580.850	8144.295	12726.592	19888.061	32592.623
Approx MoI	2035.752	3180.863	4580.442	8143.009	12723.450	19880.391	32572.035
SumMol Dif	0.000	0.000	0.000	0.000	0.000	0.000	0.000
Exact-Appr	0.080	0.197	0.407	1.287	3.142	7.670	20.588
Dif 0/00	0.039	0.062	0.089	0.158	0.247	0.386	0.632
Circle Diam 100.000							
Exact MoI	2513.355	3927.187	5655.274	10054.384	15711.105	24551.363	40232.977
Approx MoI	2513.274	3926.991	5654.867	10053.097	15707.964	24543.693	40212.387
SumMol Dif	0.000	0.000	0.000	0.000	0.000	0.000	0.000
Exact-Appr	0.081	0.196	0.407	1.287	3.142	7.670	20.590
Dif 0/00	0.032	0.050	0.072	0.128	0.200	0.312	0.512

Continues in next page...

...Continues from previous page

Circle Diam 140.000							
Exact Mol	4926.098	7697.099	11083.947	19705.357	30790.750	48113.309	78836.867
Approx Mol	4926.018	7696.902	11083.540	19704.070	30787.609	48105.637	78816.281
SumMol Dif	0.000	0.000	0.000	0.000	0.000	0.000	0.000
Exact-Appr	0.081	0.196	0.407	1.287	3.141	7.672	20.586
Dif 0/00	0.016	0.026	0.037	0.065	0.102	0.159	0.261
Circle Diam 150.000							
Exact Mol	5654.947	8835.926	12723.858	22620.756	35346.059	55230.980	90498.461
Approx Mol	5654.867	8835.729	12723.451	22619.469	35342.918	55223.309	90477.875
SumMol Dif	0.000	0.000	0.000	0.000	0.000	0.000	0.000
Exact-Appr	0.080	0.196	0.407	1.287	3.141	7.672	20.586
Dif 0/00	0.014	0.022	0.032	0.057	0.089	0.139	0.227
Circle Diam 200.000							
Exact Mol	10053.178	15708.160	22619.877	40213.676	62834.996	98182.445	160870.141
Approx Mol	10053.097	15707.964	22619.469	40212.387	62831.855	98174.773	160849.547
SumMol Dif	0.000	0.000	0.000	0.000	0.000	0.000	0.000
Exact-Appr	0.081	0.196	0.408	1.289	3.141	7.672	20.594
Dif 0/00	0.008	0.012	0.018	0.032	0.050	0.078	0.128

Each column shows a particle diameter (in cm).
Rows show:
- Area: Particle's area.
- Mol: Particle's moment of inertia.
- Exact Mol: Moment of inertia calculated with exact formula [b]=[5].
- Approx Mol: Moment of inertia calculated with approximate formula [a]=[6].
- SumMol Dif: Difference between "Exact Mol" and the value calculated as sum of each particle's Mol.
- Exact-Appr: Difference between "Exact Mol" and "Approx Mol"
- Dif 0/00: Error ratio=(Exact-Appr)/ApproxMI, in ‰ (not in %).

Mechanical Symmetry

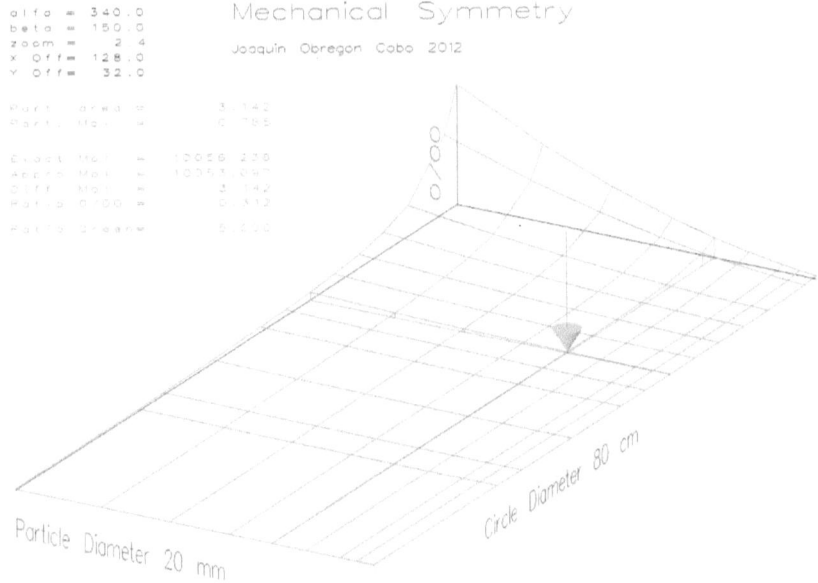

Fig. 37 MoI Graphical Interactive Comparison (B&W version)

Ap. 2.5 MoI Graphical Interactive Comparison

This program shows the difference between MoI calculated by several methods, for some values of particle radius and of the circle radius. Particle's system is a set of sixteen circles evenly located along a circumference

Particle's diameters get the values 8, 10, 12, 16, 20, 25, 32, 40, and 50—values that are, in millimeters, the usual diameters for steel reinforcing bars (rebars) for concrete. The circles where particles are located get the following values for its diameter: 20, 25, 40, 50, 60, 70, 80, 90, 100, 120, 140, 150, and 200; those values, in centimeters, are the usual values for reinforced concrete columns.

Pressing different keys:
- Up and Down arrows change circle's diameter.
- Left and Right arrows particle's diameter.
- Control + Arrows move the image on the screen.
- Shift + Arrows rotate the image on the screen.
- Z,z is Zoom.
- Control + Z changes the acceptable precision (displayed in green).
- R resets to initial view.
- B shows black background
- W shows white background

Mechanical Symmetry

PartMoI_GL.c

```c
/************************************************************
    CopyRight Joaquin Obregon Cobo 2012
    Mechanical Symmetry
    Sample program to show accuracy of approx. Formula
    Disclaimer: Use at you own risk. Author does not accept
    any responsability from the use of this code.
************************************************************/

// Includes
#include <windows.h>
#include <gl/gl.h>
#include <math.h>

// Defines
#define SAMPLES 16
#define SAMPLESf 16.0f
#define NPR 9   // Number of Particle's Radius
#define NCR 13  // Number of Circle's Radius
// This "function" to clean a little source code
#define BoFontPrint( fmt , val ) \
    glTranslated( 0 , -1.5 , 0.0 ); \
    sprintf( strinfo , fmt , val ); \
    BoFontString(strinfo);

// Function Declarations
LRESULT CALLBACK WndProc (HWND hWnd, UINT message,
WPARAM wParam, LPARAM lParam);
void EnableOpenGL (HWND hWnd, HDC *hDC, HGLRC *hRC);
void DisableOpenGL (HWND hWnd, HDC hDC, HGLRC hRC);
// Stroke Font handling
void BoFontString(const char str[]);
void BoFontFixStep(void);
void BoFontVarStep(void);
void BoFontInit(void);

// Globals
// To animate~change perspective
float alfa = 340.0;
float beta = 150.0;
float zoom =   2.4;
float xOff = 128.0;
float yOff =  32.0;
// Cursor coordinates
int activeParticle = 4;
int activeCircle = 6;
// For Calculations
const int k=4;
static float errRatioLimit = 5.0f;
static float ratios[NPR][NCR];
static float exact[NPR][NCR];
static float approx[NPR][NCR];
static float anarray[NPR][NCR];
static float inarray[NPR][NCR];
static float diffarray[NPR][NCR];
static float maxX, maxY, maxZ;
```

```
static float minX, minY, minZ;
// We are working in cm
static float
particleRadius[NPR]={0.4,0.5,0.6,0.8,1.0,1.25,1.6,2.0,2.5};
static float
circleRadius[NCR]={10,12.5,20,25,30,35,40,45,50,60,70,75,100};
// To allow black or white background choice
static float black = 0.0;
static float white = 1.0;

// Fill in an array with values
void fillTable(void)
{
  float an, In, Ik, IkExact, IkSum;
  int p,q;
  // Init maximums
  maxX = particleRadius[NPR-1]*20.0;
  minX = particleRadius[0]*20.0;
  maxY = circleRadius[NCR-1];
  minY = circleRadius[0];
  maxZ = 0.0;
  minZ = 100000000000000.0;
  // Variations on circle radius
  for ( q=0 ; q<NCR ; q++ ) {
    float cr = circleRadius [q];
    // For each Circle Radius we calculate accuracy ratio 0/00
    for ( p=0 ; p<NPR ; p++ ) {
      float pr = particleRadius[p];
      an = M_PI *pr*pr;
      In = an *pr*pr / 4;
      IkExact = k * ( In + (an * cr * cr / 2));
      Ik = k * an * cr * cr / 2;
      float diff, ratio;
      diff = IkExact - Ik;
      ratio = diff / IkExact * 1000.0;
      // Calculate max for view configuration
      pr *= 20.0;
      maxZ = maxZ < ratio ? ratio : maxZ;
      minZ = minZ > ratio ? ratio : minZ;
      // we save values to avoid recalculations
      ratios[p][q] = ratio;
      anarray[p][q] = an;
      inarray[p][q] = In;
      exact[p][q] = IkExact;
      approx[p][q] = Ik;
      diffarray[p][q] = diff;
    }
  }
}

//  Draws all data as needed
void drawTable(void)
{
  int p , q;
  float x , y , z;
  // Cursor
  // over circle
  q = activeCircle;
  y = circleRadius[q];
```

```
glColor3f (0.5f, 0.5f , 0.5f );
glBegin (GL_LINES);
glVertex3f ( minX , y , 0.0f );
glVertex3f ( maxX , y, 0.0f );
for ( p=0 ; p<NPR ; p++ ) {
  x = particleRadius[p]*20.0;
  z = ratios[p][q];
  glVertex3f ( x , y , 0.0f );
  glVertex3f ( x , y , z );
}
glEnd ();
// over particle
p = activeParticle;
x = particleRadius[p]*20.0;
glBegin (GL_LINES);
glVertex3f ( x , minY , 0.0f );
glVertex3f ( x , maxY, 0.0f );
for ( q=0 ; q<NCR ; q++ ) {
  y = circleRadius[q];
  z = ratios[p][q];
  glVertex3f ( x , y , 0.0f );
  glVertex3f ( x , y , z );
}
// Arrow (in 3D is a little long)
p = activeParticle;
x = particleRadius[p]*20.0;
q = activeCircle;
y = circleRadius[q];
z = ratios[p][q];
float arrowWidth = maxZ / 20.0;
glVertex3f ( x , y , z + maxZ );
glVertex3f ( x , y , z + arrowWidth * 4.0 );
glEnd ();
float rads;
float inc = M_PI / 128;
arrowWidth = maxZ / 20.0;
//  Arrow point circle
glColor3f (0.8f, 0.8f , 0.8f );
glBegin (GL_TRIANGLES);
for ( rads = 0.0 ; rads < 2.0 * M_PI ;  ) {
  glVertex3f ( x , y , z + arrowWidth * 4.0 );
  glVertex3f ( x + arrowWidth * cos(rads)
         , y + arrowWidth * sin(rads)
         , z + arrowWidth * 4.0 );
  rads += inc;
  glVertex3f ( x + arrowWidth * cos(rads)
         , y + arrowWidth * sin(rads)
         , z + arrowWidth * 4.0 );
}
glEnd ();
// Arrow point is a cone
glBegin (GL_TRIANGLES);
inc *= 6;
for ( rads = 0.0 ; rads < 2 * M_PI ;  ) {
  float trickcolor = 0.5 + 0.3 * sin(rads);
  glColor3f (trickcolor , trickcolor , trickcolor );
  glVertex3f ( x , y , z );
  glVertex3f ( x + arrowWidth * cos(rads)
         , y + arrowWidth * sin(rads)
```

```
                , z + arrowWidth * 4.0 );
      rads += inc;
      glVertex3f ( x + arrowWidth * cos(rads)
                 , y + arrowWidth * sin(rads)
                 , z + arrowWidth * 4.0 );
    }
    glEnd ();

    // Data Mesh
    for ( q=0 ; q<NCR ; q++ ) {
      glBegin (GL_LINE_STRIP);
      for ( p=0 ; p<NPR ; p++ ) {
        x = particleRadius[p]*20.0;
        y = circleRadius[q];
        z = ratios[p][q];
        //if ( p == activeParticle || q == activeCircle ) glColor3f (0,0,0);
        if ( z > errRatioLimit ) glColor3f (1.0f, 0.0f , 0.0f );
        else glColor3f (0.0f, 1.0f , 0.0f );
        glVertex3f ( x , y , z );
      }
      glEnd ();
    }
    for ( p=0 ; p<NPR ; p++ ) {
      glBegin (GL_LINE_STRIP);
      for ( q=0 ; q<NCR ; q++ ) {
        x = particleRadius[p]*20.0;
        y = circleRadius[q];
        z = ratios[p][q];
        //if ( p == activeParticle || q == activeCircle ) glColor3f (0,0,0);
        if ( z > errRatioLimit ) glColor3f (1.0f, 0.0f , 0.0f );
        else glColor3f (0.0f, 1.0f , 0.0f );
        glVertex3f ( x , y , z );
      }
      glEnd ();
    }
}

// WinMain
int WINAPI WinMain (HINSTANCE hInstance,
                    HINSTANCE hPrevInstance,
                    LPSTR lpCmdLine,
                    int iCmdShow)
{
    WNDCLASS wc;
    HWND hWnd;
    HDC hDC;
    HGLRC hRC;
    MSG msg;
    BOOL bQuit = FALSE;

    /* register window class */
    wc.style = CS_OWNDC;
    wc.lpfnWndProc = WndProc;
    wc.cbClsExtra = 0;
    wc.cbWndExtra = 0;
    wc.hInstance = hInstance;
    wc.hIcon = LoadIcon (NULL, IDI_APPLICATION);
```

```
  wc.hCursor = LoadCursor (NULL, IDC_ARROW);
  wc.hbrBackground = (HBRUSH) GetStockObject (BLACK_BRUSH);
  wc.lpszMenuName = NULL;
  wc.lpszClassName = "Formula Accuracy";
  RegisterClass (&wc);

  /* create main window */
  hWnd = CreateWindow (
    "Formula accuracy", "Mechanical Symmetry",
    WS_CAPTION | WS_POPUPWINDOW | WS_VISIBLE,
    // Change for a different value to get window size bigger(or smaller)
    0, 0, 1430  , 850,
    NULL, NULL, hInstance, NULL);

  /* enable OpenGL for the window */
  EnableOpenGL (hWnd, &hDC, &hRC);
  // Init Font(no drama if not initializaed, but better to do it
  BoFontInit();

  // Fill Table
  fillTable ();

  /* program main loop */
  while (!bQuit)
  {
    /* check for messages */
    if (PeekMessage (&msg, NULL, 0, 0, PM_REMOVE))
    {
      /* handle or dispatch messages */
      if (msg.message == WM_QUIT)
      {
        bQuit = TRUE;
      }
      else
      {
        TranslateMessage (&msg);
        DispatchMessage (&msg);
      }

      // Something happened ...
      // Prepare window
      // Clear
      glClearColor (black, black, black, 1.0f);
      glClear (GL_COLOR_BUFFER_BIT | GL_DEPTH_BUFFER_BIT);

      // Draw Info
      // Display information about view angles
      // Setup View
      glMatrixMode(GL_PROJECTION);
      glLoadIdentity();
      // Fefine a view with row and column characters coordinates
      glOrtho(0.0 , 70.0 , 0.0 , 64.0, -1.0, 1.0);

      char strinfo[32];

      BoFontVarStep();
      glColor3f (white, white, white);
      glPushMatrix ();
        glTranslated( 20.0 , 60.0 , 0.0 );
```

```
        glScalef( 2.0f , 2.0f , 1.0f );
      BoFontString( "Mechanical Symmetry" );
    glTranslated( 0.0 , -1.3 , 0.0 );
        glScalef( 0.7f , 0.7f , 1.0f);
      BoFontString( "Joaquin Obregon Cobo 2012" );
  glPopMatrix ();
  BoFontFixStep();
  glTranslated( 1.0 , 63.5 , 0.0 );
  BoFontPrint( "alfa = %5.1f" , alfa );
  BoFontPrint( "beta = %5.1f" , beta );
  BoFontPrint( "zoom = %5.1f" , zoom );
  BoFontPrint( "X Off= %5.1f" , xOff );
  BoFontPrint( "Y Off= %5.1f" , yOff );
  glTranslated( 0.0 , -2.0 , 0.0 );
  if ( ratios[activeParticle][activeCircle] > errRatioLimit )
     glColor3f (1.0f, 0.0f , 0.0f );
  else
     glColor3f (0.0f, 1.0f , 0.0f );
  BoFontPrint("Part. area =
%10.3f",anarray[activeParticle][activeCircle]);
  BoFontPrint( "Part. MoI   = %10.3f" ,
inarray[activeParticle][activeCircle] );
  glTranslated( 0.0 , -2.0 , 0.0 );
  BoFontPrint( "Exact MoI   = %10.3f" ,
exact[activeParticle][activeCircle] );
  BoFontPrint( "Appro MoI   = %10.3f" ,
approx[activeParticle][activeCircle] );
  BoFontPrint( "Diff  MoI   = %10.3f" ,
diffarray[activeParticle][activeCircle] );
  BoFontPrint( "Ratio 0/00  = %10.3f" ,
ratios[activeParticle][activeCircle] );
  glTranslated( 0.0 , -1.0 , 0.0 );
  BoFontPrint( "Ratio Green= %10.3f" , errRatioLimit );
  glFlush();

  // Draw content
  // Set up view
  // Set up perspective~projection
  glMatrixMode(GL_PROJECTION);
  glLoadIdentity();
  glOrtho(-2.0*maxX,2.0*maxX,-2.0*maxY,2.0*maxY,-
10.0*maxZ,10.0*maxZ);
  float distorsion = maxZ * 5 / maxX * 1024 / 728;

  glTranslatef( xOff , yOff , 0.0f );
  glRotatef (alfa, 1.0f, 0.0f, 0.0f);
  glRotatef (beta, 0.0f, 0.0f, 1.0f);
  glScalef ( zoom , zoom , zoom * 3 );

  // Draw coordinate axes
  glBegin (GL_LINES);
  glColor3f (white, white, white);
  glVertex3f(minX,minY,0.0f);glVertex3f(maxX,minY,0.0f);
  glVertex3f(maxX,minY,0.0f);glVertex3f(maxX,maxY,0.0f);
  glVertex3f(maxX,minY,0.0f);glVertex3f(maxX,minY,maxZ);
  glEnd ();

  // Draw axes labels
  // Angle label
```

```
      BoFontVarStep();
         float scale = (maxX-minX) / 25; // 25 characters
      glPushMatrix();
           glRotatef( 90, 1.0f, 0.0f , 0.0f);
           glTranslated( maxX , -1.0f , -maxY );
              glScalef( -scale, scale*distorsion , 1.0f);
              BoFontPrint( "Particle Diameter %.0f mm",
           particleRadius[activeParticle] * 20.0);
           glPopMatrix();
           // K labels
       glPushMatrix();
           glRotatef( 90, 1.0f, 0.0f , 0.0f);
           glRotatef( 90, 0.0f, 1.0f , 0.0f);
           glTranslated( maxY-(maxY-minY)/4.0 ,-1.0f ,minX);
              glScalef(-scale*1.4,scale*distorsion,1.0f);
              BoFontPrint( "Circle Diameter %.0f cm",
           circleRadius[activeCircle] * 2.0f);
           glPopMatrix();
           // Sum/k label
       glPushMatrix();
           glRotatef( 90, 0.0f, 1.0f , 0.0f);
           glRotatef( 90, 1.0f, 0.0f , 0.0f);
           glTranslated( 0.0f , maxX + 2.0f , -minY );
              glScalef(-scale*distorsion , scale , 1.0f);
           BoFontString("0/00");
           glPopMatrix();

           drawTable();

       SwapBuffers (hDC);
    }
    else
    {
      // Do nothing
    }
  }

  /* shutdown OpenGL */
  DisableOpenGL (hWnd, hDC, hRC);
  /* destroy the window explicitly */
  DestroyWindow (hWnd);
  return msg.wParam;
}


// Window Procedure to handle interaction with user and system
LRESULT CALLBACK WndProc (HWND hWnd, UINT message,
             WPARAM wParam, LPARAM lParam)
{
  static int shifted = 0;
  static int ctrled = 0;
  switch (message)
  {
  case WM_CREATE:
    return 0;
  case WM_CLOSE:
    PostQuitMessage (0);
    return 0;
```

```
case WM_DESTROY:
  return 0;

case WM_KEYUP:
  switch (wParam)
  {
  case VK_SHIFT:
    shifted = 0;
    break;
  case VK_CONTROL:
    ctrled = 0;
    break;
  }
  return 0;

case WM_KEYDOWN:
  switch (wParam)
  {
  case VK_ESCAPE:
    PostQuitMessage(0);
    break;
  case VK_SHIFT:
    shifted = 1;
    break;
  case VK_CONTROL:
    ctrled = 1;
    break;
  case 'B':
    black = 0.0f;
    white = 1.0f;
    break;
  case 'W':
    black = 1.0f;
    white = 0.0f;
    break;
  case 'R':
    alfa = 340.0;
    beta = 150.0;
    zoom =   2.4;
    xOff = 128.0;
    yOff =  32.0;
    errRatioLimit = 5.0f;
    break;
  }
  if (shifted) {
    switch (wParam)
    {
    case 'Z':
        zoom += 0.1;
      break;
    case VK_UP:
      alfa += 2;
      alfa = alfa == 360 ? 0 : alfa;
      break;
    case VK_DOWN:
      alfa -= 2;
      alfa = alfa == -2 ? 358 : alfa;
      break;
```

```
      case VK_LEFT:
        beta += 2;
        beta = beta == 360 ? 0 : beta;
        break;
      case VK_RIGHT:
        beta -= 2;
        beta = beta == -2 ? 358 : beta;
        break;
      }
    } else if (ctrled) {
      switch (wParam)
      {
      case 'Z':
        errRatioLimit += 0.1;
        errRatioLimit=errRatioLimit>10.1?0:errRatioLimit;
        break;
      case VK_UP:
        yOff += 1;
        break;
      case VK_DOWN:
        yOff -= 1;
        break;
      case VK_LEFT:
        xOff -= 1;
        break;
      case VK_RIGHT:
        xOff += 1;
        break;
      }
    } else {
      switch (wParam)
      {
      case 'Z':
         zoom -= 0.1;
         zoom = zoom < 0.1 ? 0.1 : zoom;
        break;
      case VK_UP:
        activeCircle--;
        activeCircle =activeCircle < 0 ? 0 : activeCircle;
        break;
      case VK_DOWN:
        activeCircle++;
        activeCircle=activeCircle>=NCR?NCR-1:activeCircle;
        break;
      case VK_LEFT:
        activeParticle++;
        activeParticle=activeParticle>=NPR?NPR-1:activeParticle;
        break;
      case VK_RIGHT:
        activeParticle--;
        activeParticle = activeParticle < 0 ? 0 : activeParticle;
        break;
       }
    }
    return 0;

default:
  return DefWindowProc (hWnd, message, wParam, lParam);
}
```

```c
}

// Enable OpenGL
void EnableOpenGL (HWND hWnd, HDC *hDC, HGLRC *hRC)
{
  PIXELFORMATDESCRIPTOR pfd;
  int iFormat;
  /* get the device context (DC) */
  *hDC = GetDC (hWnd);
  /* set the pixel format for the DC */
  ZeroMemory (&pfd, sizeof (pfd));
  pfd.nSize = sizeof (pfd);
  pfd.nVersion = 1;
  pfd.dwFlags = PFD_DRAW_TO_WINDOW |
    PFD_SUPPORT_OPENGL | PFD_DOUBLEBUFFER;
  pfd.iPixelType = PFD_TYPE_RGBA;
  pfd.cColorBits = 24;
  pfd.cDepthBits = 16;
  pfd.iLayerType = PFD_MAIN_PLANE;
  iFormat = ChoosePixelFormat (*hDC, &pfd);
  SetPixelFormat (*hDC, iFormat, &pfd);
  /* create and enable the render context (RC) */
  *hRC = wglCreateContext( *hDC );
  wglMakeCurrent( *hDC, *hRC );
  // Zbuffering
  //glDepthFunc(GL_LEQUAL);
  glEnable(GL_DEPTH_TEST);
  //glClearDepth(1.0);
}

// Disable OpenGL
void DisableOpenGL (HWND hWnd, HDC hDC, HGLRC hRC)
{
  wglMakeCurrent (NULL, NULL);
  wglDeleteContext (hRC);
  ReleaseDC (hWnd, hDC);
}
```

BoFont.c (Incomplete)

```c
/* Copyright (c) Mark J. Kilgard, 1994. */
/* Modified 2012 Joaquin Obregon */
/* This program is freely distributable without licensing fees
   and is provided without guarantee or warrantee expressed or
   implied. This program is -not- in the public domain. */

#if defined(_WIN32)
//#include <windows.h>
#include <gl/gl.h>
#include <stdio.h>
#pragma warning (disable:4244)   /* disable bogus conversion warnings */
#pragma warning (disable:4305)   /* VC++ 5.0 version of above warning. */
#endif

typedef struct {
  float x;
  float y;
} CoordRec, *CoordPtr;

typedef struct {
  int num_coords;
  const CoordRec *coord;
} StrokeRec, *StrokePtr;

typedef struct {
  int num_strokes;
  const StrokeRec *stroke;
  float center;
  float right;
} StrokeCharRec, *StrokeCharPtr;

typedef struct {
  const char *name;
  int num_chars;
  const StrokeCharRec *ch;
  float top;
  float bottom;
} StrokeFontRec, *StrokeFontPtr;

typedef void *GLUTstrokeFont;

#endif /* __glutstroke_h__ */

/* GENERATED FILE -- DO NOT MODIFY */
/* char: 33 '!' */
static const CoordRec char33_stroke0[] = {
  { 13.3819, 100 },
  { 13.3819, 33.3333 },
};
static const CoordRec char33_stroke1[] = {
  { 13.3819, 9.5238 },
  { 8.62, 4.7619 },
  { 13.3819, 0 },
```

```
    { 18.1438, 4.7619 },
    { 13.3819, 9.5238 },
};
static const StrokeRec char33[] = {
    { 2, char33_stroke0 },
    { 5, char33_stroke1 },
};
.
.
.
/* char: 127 */
static const CoordRec char127_stroke0[] = {
    { 52.381, 100 },
    { 14.2857, -33.3333 },
};
static const CoordRec char127_stroke1[] = {
    { 28.5714, 66.6667 },
    { 14.2857, 61.9048 },
    { 4.7619, 52.381 },
.
.
    { 52.381, 61.9048 },
    { 38.0952, 66.6667 },
    { 28.5714, 66.6667 },
};
static const StrokeRec char127[] = {
    { 2, char127_stroke0 },
    { 17, char127_stroke1 },
};
static const StrokeCharRec chars[] = {
    { 0, /* char0 */ 0, 0, 0 },
    { 0, /* char1 */ 0, 0, 0 },
    { 0, /* char2 */ 0, 0, 0 },
.
.
    { 0, /* char29 */ 0, 0, 0 },
    { 0, /* char30 */ 0, 0, 0 },
    { 0, /* char31 */ 0, 0, 0 },
    { 0, /* char32 */ 0, 52.381, 104.762 },
    { 2, char33, 13.3819, 26.6238 },
    { 2, char34, 23.0676, 51.4352 },
    { 4, char35, 36.5333, 79.4886 },
.
.
    { 3, char125, 18.7038, 41.4695 },
    { 2, char126, 45.7771, 91.2743 },
    { 2, char127, 33.3333, 66.6667 },
};
StrokeFontRec glutStrokeRoman = { "Roman", 128, chars, 119.048, -
33.3333 };

static float BoFontWid=0;
static float chrSpc=0;
static float BoFontHei=0;
static int varStep=0;// 1 paso variable - 0 paso fijo
static int centered=0;// 1 center - 0 left
```

```c
void glutStrokeInit(StrokeFontRec *font)
{
  StrokeCharRec *ch;
  const StrokeRec *stroke;
  const CoordRec *coord;
  int i, j, chr, wid;

  // Init(please, just once for Font) to simplify dependencies
  if (BoFontWid == 0) {
   for (chr=font->num_chars-1 ; chr > 0 ; chr--) {
     ch = &(font->ch[chr]);
     for (  i = ch->num_strokes,
            stroke = ch->stroke;
            i > 0; i--, stroke++) {
       wid = 0;
       for (  j = stroke->num_coords,
              coord = stroke->coord;
              j > 0; j--, coord++)
         wid = wid < coord->x ? coord->x : wid;
       ch->right = wid;
       BoFontWid = BoFontWid < wid ? wid : BoFontWid;
     }
   }
   BoFontHei=font->top - font->bottom;
  }
  chrSpc = BoFontWid/7.0f;
}

void glutStrokeCharacter(StrokeFontRec *font, int c)
{
  const StrokeCharRec *ch;
  const StrokeRec *stroke;
  const CoordRec *coord;
  int i, j, chr;
  float hal=0.0f;

  if (c < 0 || c >= font->num_chars) return;

  // Internal init(just once) to simplify dependencies
  if (BoFontWid == 0) glutStrokeInit( &glutStrokeRoman );

  ch = &(font->ch[c]);
  if (ch) {
   //  For fix / variable witdh font
   if ( !varStep )
     glTranslatef(hal=((BoFontWid - ch->right)/2.0f), 0.0, 0.0);
   for (i = ch->num_strokes, stroke = ch->stroke; i > 0; i--, stroke++) {
      glBegin(GL_LINE_STRIP);
      for (j = stroke->num_coords, coord = stroke->coord;j > 0; j--, coord++) {
        glVertex3f(coord->x , coord->y ,0.0f);
      }
      glEnd();
   }
   if ( varStep ) glTranslatef(ch->right + chrSpc, 0.0, 0.0);
   else glTranslatef(BoFontWid - hal + chrSpc, 0.0, 0.0);
  }
```

```c
}

void BoFontInit(void)
{
   glutStrokeInit( &glutStrokeRoman );
}

float BoFontStringWidth(const char str[])
{
  StrokeCharRec *ch;
  int i, j, chr , c;
  float wid=0.0f;

  if (BoFontWid == 0) BoFontInit();
  if ( varStep )
       for(chr=0 ; str[chr]!=0 ; chr++){
       c = (int) str[chr];
       if (c < 0 || c >= glutStrokeRoman.num_chars) continue;
       ch = &(glutStrokeRoman.ch[c]);
       if (ch) wid += ch->right;
       }
       else for(chr=0 ; str[chr]!=0 ; chr++) wid += BoFontWid;
       wid += chrSpc * (float)chr;
  return wid;
}

void BoFontString(const char str[])
{
  int i;
  if (BoFontWid == 0) BoFontInit();
       glPushMatrix();
       glScaled(1.0/(BoFontWid*8.0/7.0),1.0/BoFontHei,1.0);
  if (centered)
     glTranslatef( -BoFontStringWidth(str)/2.0f , 0.0, 0.0);
       for(i=0 ; str[i]!=0 ; i++)
            glutStrokeCharacter( &glutStrokeRoman , (int)str[i]);
       glPopMatrix();
}

void BoFontFixStep(void) { varStep=0; }
void BoFontVarStep(void) { varStep=1; }

void BoFontCenter(void) { centered=1; }
void BoFontLeft(void) { centered=0; }
```

Mechanical Symmetry

Appendix 3: Computer Programs for Chapter 5

Mechanical Symmetry

Mechanical Symmetry Calcula MoI

Joaquin Obregon Cobo 2012

```
MoI
(Exact-Exact)/Exact ~ 0.0%
Exact    = 0.162
(Exact-Approx)/Approx ~ 125.0%
Approx   = 0.072
(Exact-Sums)/Sums ~ 0.0%
Sums     = 0.162
(Exact-Generic)/Generic ~ 0.0%
Generic  = 0.162
(Exact-Circle)/Circle ~ 79.3%
Circle   = 0.785
```

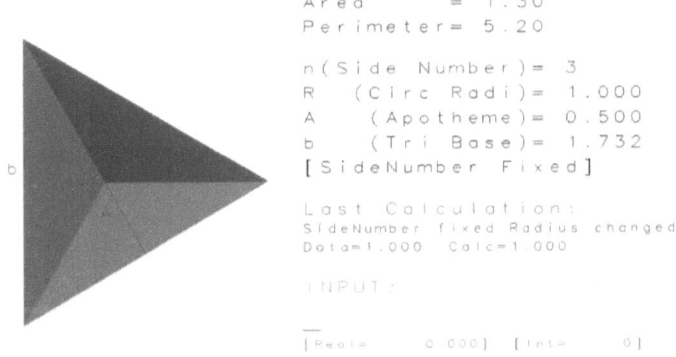

```
Area       = 1.30
Perimeter= 5.20

n(Side Number)= 3
R  (Circ Radi)= 1.000
A  (Apotheme )= 0.500
b  (Tri Base )= 1.732
[SideNumber Fixed]

Last Calculation:
SideNumber fixed Radius changed
Data=1.000  Calc=1.000

INPUT:

[Real=     0.000] [Int=    0]
```

Fig. 38 Regular Polygons MoI Interactive Calculation (B&W version)

Ap. 3.1 Regular Polygons MoI Interactive Calculation

This program solves the polygon as a function of input data and calculates its MoI by different methods, as well as its area, perimeter, and MoI for its circumcircle.

Different values for each one of the parameters can be introduced:

- Number of sides (N key)
- Circumcircle's radius (R key)
- Polygon's apothem (A key)
- Polygon's side (B key)

To do it, type the value and then press the corresponding key (For radius = 2, type 2 and then press the R key).

If you press a key for a parameter without typing a value or with zero value, this parameter becomes fixed for the next calculations.

If you input a value for a parameter yet fix the number of sides, it is considered as the second fixed parameter.

Pressing up and down arrow keys modifies the number of sides.

To change background color from black and white, press key W.

MoI_Calc.c

```
/*************************************************************
    CopyRight Joaquin Obregon Cobo 2012
    Mechanical Symmetry
    Sample program to illustrate MoI for Polígonos Regulares
    Disclaimer: Use at you own risk. Author does not accept
    any responsability from the use of this code.
*************************************************************/

// Includes
#include <windows.h>
#include <gl/gl.h>
#include <math.h>

// Defines
// This "function" to clean a little source code
#define BoFontPrint( fmt , val ) \
    glTranslated( 0 , -1.5 , 0.0 ); \
    sprintf( strinfo , fmt , val ); \
    BoFontString(strinfo);
// Function Declarations
LRESULT CALLBACK WndProc (HWND hWnd, UINT message,WPARAM wParam,
LPARAM lParam);
void EnableOpenGL (HWND hWnd, HDC *hDC, HGLRC *hRC);
void DisableOpenGL (HWND hWnd, HDC hDC, HGLRC hRC);
// Stroke Font handling
void BoFontString(const char str[]);
void BoFontFixStep(void);
void BoFontVarStep(void);
void BoFontInit(void);

// Types
typedef enum PolParam // Polygon defining parameters
{
  Radius=0, // Ex Radius
  Apotheme, // Equals Generator triangle height
  Sides,
  base, // Generator triangle basis
} polParamType;

// Globals

// Polygon definition
float R = 1.0; // Circunscr. Radius
float apo; // Apotheme
int n = 3; // Number of sides
float b,h; // Base and height of generator triangle
```

```c
polParamType blocked = Sides; // The second parameter to calculate
polygon from
char *PolParamNames[]={"Radius","Apotheme","SideNumber","Side"};

// Input/output aux var
float in_float;
#define IN_MAXFLOAT 100000.0f
int in_int;
#define IN_MAXINT 16384
#define MAX_IN 10
char in_str[MAX_IN];
char feedback_param[64], feedback_values[64];
int cursor=0;
int decimal=0; // 1--> decimal point in string , 0--> no decimal
point
#define INIT_INPUT
{in_str[0]='_';in_str[0]=0;decimal=0;cursor=0;in_float=0.0F;in_int
=0;}

// To define view
float black = 0.0f;
float white = 1.0f;

// Action

void drawRegularPolygon(float x, float y, float R, float n)
// Center (x,y)
// Cinscunsc Circunf Rad R
{
  float rads, inc, vy, vx, col, apo;
  int i;
  glBegin (GL_TRIANGLES);
  vy = y;
  vx = x + R;
  inc = 2 * M_PI / n;
  col = 0.6 / n;
  // Draw polygon as triangles
  for ( i = 1 ; i <= n ; i++ ) {
    glColor3f (0.0f, col*i/2.0 , 0.3+col*i );
    glVertex3f ( x , y , -0.1f );
    glVertex3f ( vx , vy , -0.1f );
    rads = inc * i;
    vy = y + R * sin(rads);
    vx = x + R * cos(rads);
    glVertex3f ( vx , vy , -0.1f );
  }
  glEnd ();
  // Draw Radius
  glBegin (GL_LINES);
    glColor3f (white+black/2.0 , white+black/2.0 , 0.0 );
    glVertex3f ( x , y , 0.0f );
    glVertex3f ( x+R , y , 0.0f );
  glEnd ();
  glPushMatrix ();
    glTranslatef( x+R/2.4 , y+0.5 , 0.0 );
    BoFontString( "R" );
  glPopMatrix();
  // Draw Apotheme
  glBegin (GL_LINES);
```

```
    glVertex3f ( x , y , 0.0f );
    i = (n*2)/3;
    rads = inc * (i + 0.5 );
    apo = R * cos(M_PI/n);
    vy = y + apo * sin(rads);
    vx = x + apo * cos(rads);
    glVertex3f ( vx , vy , 0.0f );
  glEnd ();
  glPushMatrix ();
    glTranslatef( (x+vx)/2.0 -1.5 , (y+vy)/2.0 , 0.0 );
    BoFontString( "A" );
  glPopMatrix();
  // Draw Side
  glBegin (GL_LINES);
    i = n/3;
    rads = inc * i;
    vy = y + R * sin(rads);
    vx = x + R * cos(rads);
    glVertex3f ( vx , vy , 0.0f );
    i++;
    rads = inc * i;
    y = y + R * sin(rads);
    x = x + R * cos(rads);
    glVertex3f ( x , y , 0.0f );
  glEnd ();
  glPushMatrix ();
    glTranslatef( (x+vx)/2.0 -1.0 , (y+vy)/2.0 +0.5 , 0.0 );
    BoFontString( "b" );
  glPopMatrix();
}

// Calculations to Solve the Polygon
int polyCalc( float param , polParamType paramType)
// Not very optimized algorithm but quite easy to follow
// Returns 0 if there was an ERROR, 1 if everything OK
{
  float _param = param;
  sprintf( feedback_param , "Calculation not feasible" );
  sprintf( feedback_values , "Incompatible Data & Constraints" );
  if ( n < 3 || param == 0 ) return( 0 ); // No zero div
  if ( paramType == Sides && param < 3.0 ) return( 0 ); // TRI minimum
  switch( blocked ){
      case Radius:
          if ( R == 0.0 ) return( 0 ); // No zero div
          switch( paramType ){
              case Radius:
                  R = param;
                  apo = R * cos(M_PI / (float) n);
                  b = 2 * R * sin(M_PI / (float) n);
                  break;
              case Sides:
                  n = param;
                  apo = R * cos(M_PI / (float) n);
                  b = 2 * R * sin(M_PI / (float) n);
                  break;
              case Apotheme:
                  if (param >= R)return(0);// Geometry
                  // float comparison is quiet tricky
```

```
                if(param+0.001<R*cos(M_PI/3)) return(0);
                apo = param;
                n = (int) (M_PI / acos(apo/R)+0.5);
                b = 2 * R * sin(M_PI / (float) n);
                _param=apo=R*cos(M_PI / (float) n);
                break;
            case base:
                if (param >= 2 * R * sin(M_PI/3.0))
return( 0 ); // Geometry
                b = param;
                n = (int) (M_PI/asin(b/(2*R))+0.5);
                _param=b=2*R*sin(M_PI / (float) n);
                apo = R * cos(M_PI / (float) n);
                break;
            default:
                return(0);// Error
        }
        break;
    case Apotheme:
        switch( paramType ){
            case Radius:
                if ( R == 0.0 ) return( 0 );
// No zero div
                if (apo >= param) return( 0 );
// Geometry
                if (param > apo/cos(M_PI/3)) return(0); // Geo
                R = param;
                n = (M_PI / acos(apo/R))+0.5;
                _param = R = apo / cos(M_PI / (float) n);
                b = 2 * apo * tan(M_PI / (float) n);
                break;
            case Apotheme:
                apo = param;
                R = apo / cos(M_PI / (float) n);
                b = 2 * apo * tan(M_PI / (float) n);
                break;
            case Sides:
                n = param;
                R = apo / cos(M_PI / (float) n);
                b = 2 * apo * tan(M_PI / (float) n);
                break;
            case base:
                if (param>=2*apo*tan(M_PI/3)) // Geo
                    return( 0);
                b = param;
                if ( apo == 0.0 ) return( 0 );
// No zero div
                n = M_PI / atan(b/(2*apo))+0.5;
                _param=b=2*apo*tan(M_PI/(float) n);
                R = apo / cos(M_PI / (float) n);
                break;
            default:
                return(0);// Error
        }
        break;
    case Sides:
        switch( paramType ){
            case Radius:
                R = param;
```

```
                b = 2 * R * sin(M_PI / (float) n);
                apo = R * cos(M_PI / (float) n);
                break;
            case Sides:
                n = param;
                b = 2 * R * sin(M_PI / (float) n);
                apo = R * cos(M_PI / (float) n);
                break;
            case Apotheme:
                apo = param;
                R = apo / cos(M_PI / (float) n);
                b = 2 * R * sin(M_PI / (float) n);
                break;
            case base:
                b = param;
                R = b / (2* sin(M_PI / (float) n) );
                apo = R * cos(M_PI / (float) n);
                break;
            default:
                return(0);// Error
        }
        break;
    case base:
        switch( paramType ){
            case Radius:
                if ( param == 0.0 ) return( 0 );
// No zero div
                if ( param <= b/(2*sin(M_PI/3)))
// GEometry
                    return( 0 );
                R = param;
                n = M_PI / asin(b/(2*R))+0.5;
                _param=R=b/(2*sin(M_PI/(float) n));
                apo=b/(2 * tan(M_PI / (float) n));
                break;
            case Apotheme:
                if ( param == 0.0 ) return( 0 );
// No zero div
                if ( param <= b/(2*tan(M_PI/3)) ) //Geometry
                    return( 0 );
                apo = param;
                n = M_PI / atan(b/(2*apo))+0.5;
                apo=b/(2 * tan(M_PI / (float) n));
                _param=R=b/(2*sin(M_PI/(float) n));
                break;
            case base:
                b = param;
                R = b / (2 * sin(M_PI / (float) n));
                apo=b / (2 * tan(M_PI / (float) n));
                break;
            case Sides:
                n = param;
                R = b / (2 * sin(M_PI / (float) n));
                apo=b / (2 * tan(M_PI / (float) n));
                break;
            default:
                return(0);// Error
        }
        if ( n == 0 ) return( 0 ); // No zero div
```

```
        //b = 2 * R * sin(M_PI / (float) n);
        //b = 2 * apo * tan(M_PI / (float) n);
        break;
      default:
        return(0);// Error
  }
  // Fill in the feedback with corresponding information
  sprintf( feedback_param , "%s fixed %s changed ",
              PolParamNames[blocked] , PolParamNames[paramType]);
  sprintf( feedback_values , "Data=%.3f  Calc=%.3f", param ,
_param );
  return(1);
}


float poly_area () {
    return n * b * apo / 2.0;
}

float poly_moi_Exact () { // Formula from Mechanical Symmetry Book
            // (b*h*(12*h^2+b^2)*k)/96
    return( (float)n * b * apo / 96 *(b*b + 12 * apo*apo) );
}

float poly_moi_Approx () { // Formula from Mechanical Symmetry
Book
              // (b*h^3*k)/9
    return( n * b * apo*apo*apo / 9 );
}

float poly_moi_sums () { // Sum of triangles MoI. Rotation +
Traslation
    int i;
    float rads, inc, // aux to iterate the polygon
      dist, // Distance from axis to the rotated center of mass
      Ixx; // What can this be?
    // Principal moments
    float Iu = (b*b*b * apo) / 48;
    float Iv = (b * apo*apo*apo) / 36;
    // Area for the basic triangle
    float area = b * apo / 2;
    // Init iteration
    rads = (inc = 2.0 * M_PI / (float)n) / 2.0;
    Ixx = 0.0;
    // Process all triangles
    for ( i = 0 ;  i < n ;  i++ ) {
                // Distance from axis to the rotated center of
mass
      dist = apo * sin(rads) * 2.0 / 3.0;
      // Moment with the rotation and traslation
      float moi = Iu * cos(rads)*cos(rads) + Iv *
sin(rads)*sin(rads);
      float steiner = area * dist*dist;
      Ixx += moi + steiner;
      rads += inc;
    }
    return( Ixx );
}
```

```
float poly_moi_generic () { // Generic formula for CLOSED polygons
  int i;
  float rads, inc, // aux to iterate the polygon
     x0,y0,x1,y1, // coordinates of two consecutive vertex
     Ixx; // What can this be?
  // Init iteration
  inc = 2.0 * M_PI / (float)n;
  Ixx = 0.0;
  // Process all vertex  (  order change sign  )
  for ( i = 0 ;  i < n ;) {
    rads = inc * i++;
    x0 = R * cos(rads);
    y0 = R * sin(rads);
    rads = inc * i;
    x1 = R * cos(rads);
    y1 = R * sin(rads);
    Ixx -= (x1-x0)*(y1+y0)*(y1*y1+y0*y0);
  }
  return( Ixx / 12.0 );
}


// WinMain
int WINAPI WinMain (HINSTANCE hInstance,
          HINSTANCE hPrevInstance,
          LPSTR lpCmdLine,
          int iCmdShow)
{
  WNDCLASS wc;
  HWND hWnd;
  HDC hDC;
  HGLRC hRC;
  MSG msg;
  BOOL bQuit = FALSE;
  float cubo = 0.0;
  char strinfo[32];// We use it with BoFontPrint macro


  /* register window class */
  wc.style = CS_OWNDC;
  wc.lpfnWndProc = WndProc;
  wc.cbClsExtra = 0;
  wc.cbWndExtra = 0;
  wc.hInstance = hInstance;
  wc.hIcon = LoadIcon (NULL, IDI_APPLICATION);
  wc.hCursor = LoadCursor (NULL, IDC_ARROW);
  wc.hbrBackground = (HBRUSH) GetStockObject (BLACK_BRUSH);
  wc.lpszMenuName = NULL;
  wc.lpszClassName = "steiner";
  RegisterClass (&wc);

  /* create main window */
  hWnd = CreateWindow (
    "Steiner", "Mechanical Symmetry",
    WS_CAPTION | WS_POPUPWINDOW | WS_VISIBLE,
    0, 0, 960, 960 ,
    NULL, NULL, hInstance, NULL);

  /* enable OpenGL for the window */
```

```
    EnableOpenGL (hWnd, &hDC, &hRC);
    // Init Font
    BoFontInit();

    // Init
    INIT_INPUT
    n=3;
    polyCalc( 1.0 , Radius );

    /* program main loop */
    while (!bQuit)
    {
      /* check for messages */
      if (PeekMessage (&msg, NULL, 0, 0, PM_REMOVE))
      {
        /* handle or dispatch messages */
        if (msg.message == WM_QUIT)
        {
          bQuit = TRUE;
        }
        else
        {
          TranslateMessage (&msg);
          DispatchMessage (&msg);
        }
        // Something happened ...
        // Prepare window
        // Clear
        glClearColor (black, black, black, 1.0f);
        glClear (GL_COLOR_BUFFER_BIT | GL_DEPTH_BUFFER_BIT);

        // Draw content
        // Set up view
        // Set up perspective~projection
        glMatrixMode(GL_PROJECTION);
        glLoadIdentity();

        // Draw Info
        // Display information
        // Setup View
        glClear ( GL_DEPTH_BUFFER_BIT);
        glMatrixMode(GL_PROJECTION);
        glLoadIdentity();
        // We define a view with row and column characters
coordinates
        glOrtho(0.0 , 50.0 , 0.0 , 50.0, -1.0 , 1.0 );

        BoFontVarStep();
        glColor3f (white, white, white);
        glPushMatrix ();
          glTranslatef( 1.0 , 46 , 0.0 );
             glScalef( 2.5f , 2.5f , 1.0f);
            BoFontString( "Mechanical Symmetry Calcula MoI" );
          glTranslatef( 0.0 , -1.3 , 0.0 );
             glScalef( 0.5f , 0.5f , 1.0f);
            BoFontString( "Joaquin Obregon Cobo 2012" );
        glPopMatrix ();
        // Display DATA
        BoFontFixStep();
```

```
glPushMatrix ();
  glTranslatef( 23.0 , 23.0 , 0.0 );
  BoFontPrint( "Area     = %.2f" , poly_area() );
  BoFontPrint( "Perimeter= %.2f" , n*b );
  glTranslatef( 0.0 , -1.0 , 0.0 );
  BoFontPrint( "n(Side Number)= %.1d" , n );
  BoFontPrint( "R  (Circ Radi)= %.3f" , R );
  BoFontPrint( "A   (Apotheme)= %.3f" , apo );
  BoFontPrint( "b   (Tri Base)= %.3f" , b );
  BoFontPrint( "[%s Fixed]" , PolParamNames[blocked] );
  glTranslatef( 0.0 , -1.0 , 0.0 );
  glColor3f (white, white , black/2.0 );
  BoFontPrint("Last Calculation:", in_int);
  glScalef( 1/1.35 , 1/1.35 , 1.0 );
  BoFontPrint( "%-s" , feedback_param );
  BoFontPrint( "%-s" , feedback_values );
  glScalef( 1.35 , 1.35 , 1.0 );
  glTranslatef( 0.0 , -1.0 , 0.0 );
  glColor3f (0.0, 1.0 , 0.0 );
  BoFontPrint("INPUT:", in_int);
  glScalef( 1.35 , 1.35 , 1.0 );
  BoFontPrint( "%-s_" , in_str );
  glTranslatef( 0.0 , -1.0 , 0.0 );
  glScalef( 0.5 , 0.5 , 1.0 );
  sprintf(strinfo,
      "[Real=%10.3f]   [Int=%6.1d]",in_float, in_int );
  BoFontString( strinfo );
glPopMatrix ();
glPushMatrix ();
  // Print MoI values
  glColor3f (white, white , white );
  glTranslatef( 2.0 , 42.0 , 0.0 );
  glScalef( 1.5 , 1.5 , 1.0 );
  BoFontPrint( "M o I" , 0 );
  glTranslatef( 0.0 , -0.50 , 0.0 );
  float Iex = poly_moi_Exact();
  float Iap =  poly_moi_Approx();
  float Isu =  poly_moi_sums();
  float Ige =  poly_moi_generic();
  float Ici =  M_PI * R*R*R*R / 4.0;
  BoFontPrint( "Exact   = %.3f" , Iex );
  glTranslatef( 0.0 , -0.250 , 0.0 );
  BoFontPrint( "Approx = %.3f" , Iap );
  glTranslatef( 0.0 , -0.250 , 0.0 );
  BoFontPrint( "Sums    = %.3f" , Isu );
  glTranslatef( 0.0 , -0.250 , 0.0 );
  BoFontPrint( "Generic= %.3f" , Ige );
  glTranslatef( 0.0 , -1.0 , 0.0 );
  BoFontPrint( "Circle = %.3f" , Ici );
glPopMatrix ();
BoFontVarStep();
glPushMatrix ();
  // Print deviations/errors
  glColor3f (0.4, 0.4 , 0.4 );
  glTranslatef( 2 , 41.7 , 0.0 );
  glScalef( 1.5/2 , 1.5/2 , 1.0 );
  BoFontPrint( " " , 0 );
  BoFontPrint( " " , 0 );
  glTranslatef( 0.0 , -0.40 , 0.0 );
```

```
        BoFontPrint( "(Exact-Exact)/Exact ~ %.1f%%" , 0.0 );
        BoFontPrint( "" /*" %.3f"*/ , 0.0 );
        glTranslatef( 0.0 , -0.50 , 0.0 );
        BoFontPrint( "(Exact-Approx)/Approx ~ %.1f%%",fabs(Iex-
Iap)/Iap*100.0);
        BoFontPrint( "" /*" %.3f"*/ , (Iex-Iap) );
        glTranslatef( 0.0 , -0.50 , 0.0 );
        BoFontPrint( "(Exact-Sums)/Sums ~ %.1f%%" , fabs(Iex-
Isu)/Isu * 100.0 );
        BoFontPrint( "" /*" %.3f"*/ , (Iex-Isu) );
        glTranslatef( 0.0 , -0.50 , 0.0 );
        BoFontPrint( "(Exact-Generic)/Generic ~ %.1f%%",fabs(Iex-
Ige)/Ige*100.0);
        BoFontPrint( "" /*" %.3f"*/ , (Iex-Ige) );
        glTranslatef( 0.0 , -2.0 , 0.0 );
        BoFontPrint( "(Exact-Circle)/Circle ~ %.1f%%", fabs(Iex-
Ici)/Ici*100.0 );
        BoFontPrint( "" /*" %.3f"*/ , (Iex-Ici) );
      glPopMatrix ();

      drawRegularPolygon( 11.0 , 11.0 , 10 , n );

      glFlush();

      SwapBuffers(hDC);
    }
    else
    {
      // nothing to do
    }
  }

  /* shutdown OpenGL */
  DisableOpenGL (hWnd, hDC, hRC);
  /* destroy the window explicitly */
  DestroyWindow (hWnd);
  return msg.wParam;
}

/******************* Window Procedure *******************/
LRESULT CALLBACK WndProc (HWND hWnd, UINT message,
            WPARAM wParam, LPARAM lParam)
{
  static int shifted = 0;
  static int ctrled = 0;
  switch (message)
  {
  case WM_CREATE:
    return 0;
  case WM_CLOSE:
    PostQuitMessage (0);
    return 0;
  case WM_DESTROY:
    return 0;
  case WM_KEYDOWN:
    switch (wParam)
    {
      break;
    case '0':
```

```
        case '1':
        case '2':
        case '3':
        case '4':
        case '5':
        case '6':
        case '7':
        case '8':
        case '9':
           if (cursor < MAX_IN ) in_str[cursor++]= wParam;
           in_str[cursor]='\0';
           break;
        case VK_NUMPAD0:
        case VK_NUMPAD1:
        case VK_NUMPAD2:
        case VK_NUMPAD3:
        case VK_NUMPAD4:
        case VK_NUMPAD5:
        case VK_NUMPAD6:
        case VK_NUMPAD7:
        case VK_NUMPAD8:
        case VK_NUMPAD9:
            if(cursor<MAX_IN)in_str[cursor++]='0'+wParam-VK_NUMPAD0;
           in_str[cursor]='\0';
           break;
        case VK_DECIMAL:
        case 0xbe:
           if (cursor<MAX_IN&&!decimal)in_str[cursor++]= '.';
           in_str[cursor]='\0';
           decimal |= 1;
           break;
        case ',':
           break;
        case VK_BACK:
           if (cursor > 0) {
                decimal &= (in_str[--cursor] != '.');
                in_str[cursor]='\0';
                if ( cursor == 0 ) in_float=in_int=0;
           }
           break;
        case 'W':
          if ( white ) { black = 1.0f; white = 0.0f; }
          else { black = 0.0f; white = 1.0f; }
          break;
        case 'R':
           if( in_float == 0.0 ) blocked = Radius;
           else if ( polyCalc( in_float , Radius ) ) INIT_INPUT;
          break;
        case 'N':
           if( in_float == 0.0 ) blocked = Sides;
           else if ( polyCalc( in_float , Sides ) ) INIT_INPUT;
          break;
        case 'B':
           if( in_float == 0.0 ) blocked = base;
           else if ( polyCalc( in_float , base ) ) INIT_INPUT;
          break;
        case 'A':
           if( in_float == 0.0 ) blocked = Apotheme;
           else if (polyCalc(in_float , Apotheme ) ) INIT_INPUT;
```

```
        break;
    case VK_UP:
      if ( n < IN_MAXINT ) polyCalc( (float) ++n , Sides );
        break;
    case VK_DOWN:
      if ( n > 3 ) polyCalc( (float) --n , Sides );
        break;
    case VK_LEFT:
        break;
    case VK_RIGHT:
        break;
    }
    sscanf( in_str, "%f" , &in_float );
    if ( in_float > IN_MAXFLOAT ) in_float = IN_MAXFLOAT;
    sscanf( in_str, "%d" , &in_int );
    if (in_int > IN_MAXINT || in_int < 0 ) in_int = IN_MAXINT;
    return 0;
  case WM_KEYUP:
    switch (wParam)
    {
    case VK_ESCAPE:
      PostQuitMessage(0);
      break;
    }
    return 0;
  default:
    return DefWindowProc (hWnd, message, wParam, lParam);
  }
}

// Enable OpenGL
void EnableOpenGL (HWND hWnd, HDC *hDC, HGLRC *hRC)
{
  PIXELFORMATDESCRIPTOR pfd;
  int iFormat;
  /* get the device context (DC) */
  *hDC = GetDC (hWnd);
  /* set the pixel format for the DC */
  ZeroMemory (&pfd, sizeof (pfd));
  pfd.nSize = sizeof (pfd);
  pfd.nVersion = 1;
  pfd.dwFlags = PFD_DRAW_TO_WINDOW |
    PFD_SUPPORT_OPENGL | PFD_DOUBLEBUFFER;
  pfd.iPixelType = PFD_TYPE_RGBA;
  pfd.cColorBits = 24;
  pfd.cDepthBits = 16;
  pfd.iLayerType = PFD_MAIN_PLANE;
  iFormat = ChoosePixelFormat (*hDC, &pfd);
  SetPixelFormat (*hDC, iFormat, &pfd);
  /* create and enable the render context (RC) */
  *hRC = wglCreateContext( *hDC );
  wglMakeCurrent( *hDC, *hRC );
  // Zbuffering
  glEnable(GL_DEPTH_TEST);
}

// Disable OpenGL
void DisableOpenGL (HWND hWnd, HDC hDC, HGLRC hRC)
{
```

```
    wglMakeCurrent (NULL, NULL);
    wglDeleteContext (hRC);
    ReleaseDC (hWnd, hDC);
}
```

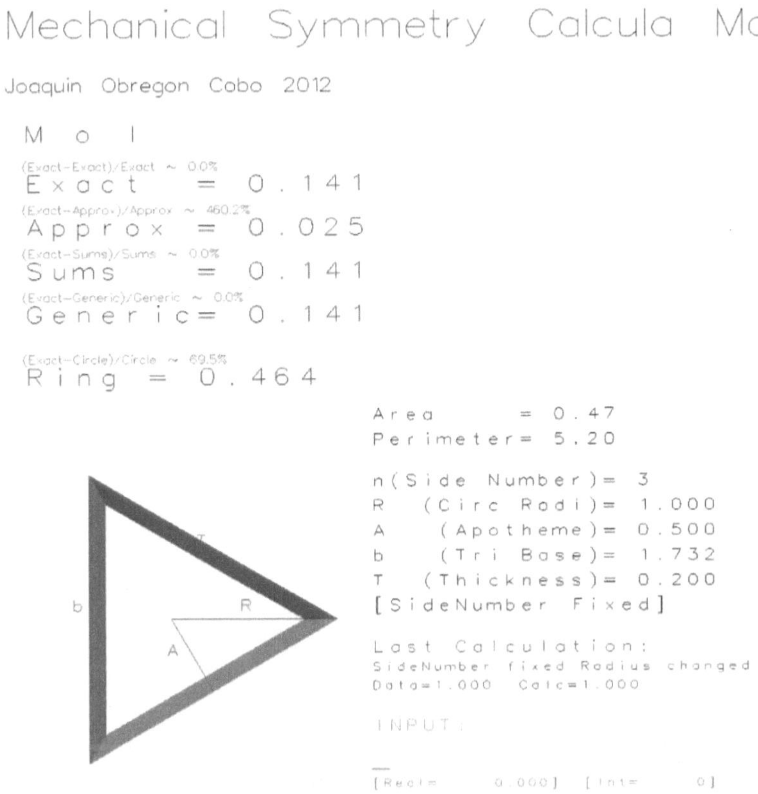

Mechanical Symmetry Calcula MoI

Joaquin Obregon Cobo 2012

```
M  o  I
(Exact-Exact)/Exact ~ 0.0%
E x a c t    =  0 . 1 4 1
(Exact-Approx.)/Approx ~ 460.2%
A p p r o x  =  0 . 0 2 5
(Exact-Sums)/Sums ~ 0.0%
S u m s      =  0 . 1 4 1
(Exact-Generic)/Generic ~ 0.0%
G e n e r i c=  0 . 1 4 1

(Exact-Circle)/Circle ~ 69.5%
R i n g   =   0 . 4 6 4
```

```
Area       = 0.47
Perimeter= 5.20

n(Side Number)= 3
R   (Circ Radi)= 1.000
A   (Apotheme)= 0.500
b   (Tri Base)= 1.732
T   (Thickness)= 0.200
[SideNumber Fixed]

Last Calculation:
SideNumber fixed Radius changed
Data=1.000  Calc=1.000

INPUT:

[Real= 0.000]  [Int= 0]
```

Fig. 39 Regular Polygonal Tubes MoI Interactive Calculation (B&W version)

Ap. 3.2 Regular Polygonal Tubes MoI Interactive Calculation

This program solves the polygon as a function of input data and calculates the MoI for a tube with thickness t by several methods, as well as its area, perimeter, and MoI for a corresponding circumcircle.

Different values for each one of the parameters can be introduced:
- Number of sides (N key)
- Circumcircle's radius (R key)
- Polygon's apothem (A key)
- Polygon's side (B key)
- Tube's thickness (T key)

To do it, type the value and then press the corresponding key (to give the radius a value of 2, type 2 and then press the R key).

If you press a key for a parameter without typing a value or with zero value, this parameter becomes fixed for the next calculation.

Mechanical Symmetry

If you input a value for a parameter yet fix the number of sides, it is considered as the second fixed parameter.

Pressing up and down arrow keys modifies the number of sides.

To change background color from black and white, press key W.

                           Excerpt from MoI_Calc.c

```
/************************************************************
    CopyRight Joaquin Obregon Cobo 2012
    Mechanical Symmetry
    Sample program to illustrate MoI for Poligonos Regulares
    Disclaimer: Use at you own risk. Author does not accept
    any responsability from the use of this code.
************************************************************/
float poly_moi_Exact () { // Formula from Mechanical Symmetry Book
// -(b2*(12*h2^2+b2^2)*k*t*(t-2*h2)*(t^2-2*h2*t+2*h2^2))/(96*h2^3)
    float h = apo;
    return( -(b*(12*h*h+b*b)*n*t*(t-2*h)*(t*t-
2*h*t+2*h*h))/(96*h*h*h) );
}
float poly_moi_Approx () { // Formula from Mechanical Symmetry Book
// -(b2*k*t*(t-2*h2)*(t((2*t-3*h2)/(3*(t-2*h2)))-h2)^2)/(4*h2)
    float h = apo;
    float aux = (t*((2*t-3*apo)/(3*(t-2*apo)))-apo);
    return( -(b*n*t*(t-2*apo)* aux*aux)/(4*apo) );
}
float poly_moi_sums () { // Sum of triangles MoI. Rotation +
Traslation
    int i;
    float rads, inc, // aux to iterate the polygon
        dist, //Dist from axis to the rotated center of mass
        apuco, // An smaller Apotheme
        bit, // An Smaller b
        Ixx; // What can this be?
    // Outer polygon
    // Principal moments
    float Iu = (b*b*b * apo) / 48;
    float Iv = (b * apo*apo) / 36;
    // Area for the basic triangle
    float area = b * apo / 2;
    // Init iteration
    rads = (inc = 2.0 * M_PI / (float)n) / 2.0;
    Ixx = 0.0;
    // Process all triangles
    for ( i = 0 ;  i < n ; i++ ) {
            // Distance from axis to the rotated center of mass
        dist = apo * sin(rads) * 2.0 / 3.0;
        // Moment with the rotation and traslation
        float moi = Iu * cos(rads)*cos(rads) + Iv *
sin(rads)*sin(rads);
        float steiner = area * dist*dist;
        Ixx += moi + steiner;
```

```
        rads += inc;
    }
    // Inner polygon
    apuco = apo - t;
    bit = b * (apo-t)/apo;
    // Principal moments
    Iu = (bit*bit*bit * apuco) / 48;
    Iv = (bit * apuco*apuco*apuco) / 36;
    // Area for the basic triangle
    area = bit * apuco / 2;
    // Init iteration
    rads = (inc = 2.0 * M_PI / (float)n) / 2.0;
    for ( i = 0 ;  i < n ; i++ ) {
        // Distance from axis to the rotated center of mass
        dist = apuco * sin(rads) * 2.0 / 3.0;
        // Moment with the rotation and traslation
        float moi = Iu * cos(rads)*cos(rads) + Iv *
sin(rads)*sin(rads);
        float steiner = area * dist*dist;
        Ixx -= moi + steiner;
        rads += inc;
    }
    return( Ixx );
}

float poly_moi_generic () { // Generic formula for CLOSED polygons
  int i;
  float rads, inc, // aux to iterate the polygon
        x0,y0,x1,y1, // coordinates of two consecutive vertex
        r, // Inner radius
        Ixx; // What can this be?
  // Init iteration
  inc = 2.0 * M_PI / (float)n;
  Ixx = 0.0;
  // Process all vertex   ( order change sign  )
  // Outer polygon
  for ( i = 0 ;  i < n ;) {
    rads = inc * i++;
    x0 = R * cos(rads);
    y0 = R * sin(rads);
    rads = inc * i;
    x1 = R * cos(rads);
    y1 = R * sin(rads);
    Ixx -= (x1-x0)*(y1+y0)*(y1*y1+y0*y0);
  }
  // Inner polygon
  r = R * (apo-t)/apo;
  for ( i = 0 ;  i < n ;) {
    rads = inc * i++;
    x0 = r * cos(rads);
    y0 = r * sin(rads);
    rads = inc * i;
    x1 = r * cos(rads);
    y1 = r * sin(rads);
    Ixx += (x1-x0)*(y1+y0)*(y1*y1+y0*y0);
  }
  return( Ixx / 12.0 );
}
```

Mechanical Symmetry

Fig. 40 Regular Polygon Based Stars MoI Interactive Calculation (B&W version)

Ap. 3.3 Regular Polygon Based Stars MoI Interactive Calculation

This program solves the polygon as a function of input data and calculates the MoI for a star by different methods, as well as its area, perimeter, and MoI for its circumcircle.

The star is formed by adding a triangle opposed by the polygon's side and equal to the one formed by the two vertexes of the side and the polygon's center.

Different values for each one of the parameters can be introduced:
- Number of sides (N key)
- Circumcircle's radius (R key)
- Polygon's apothem (A key)
- Polygon´s side (B key)

To do it, type the value and then press the corresponding key (to give the radius a value of 4, type 4 and then press the R key).

If you press a key for a parameter without typing a value or with zero value, this parameter becomes fixed for the calculation following.

If you input a value for a parameter yet fix the number of sides, it is considered the second fixed parameter.

Pressing up and down arrow keys modifies the number of sides.

To change background color from black and white, press key W.

Excerpt from MoI_Calc.c

```
/*************************************************************
    CopyRight Joaquin Obregon Cobo 2012
    Mechanical Symmetry
    Sample program to illustrate MoI for Polígonos Regulares

    Disclaimer: Use at you own risk. Author does not accept
    any responsability from the use of this code.
*************************************************************/
float poly_moi_Exact () { // Formula from Mechanical Symmetry Book
    // (b*h*(28*h^2+b^2)*k)/48
    return ( b * apo * ( 28.0 * apo*apo + b*b ) * n) / 48.0;
}
float poly_moi_Approx () { // Formula from Mechanical Symmetry Book
                // 0.5*b*h^3*k
    return 0.5 * b * apo*apo*apo *n;
}
float poly_moi_sums () { // Sum of triangles MoI. Rotation + Traslation
    int i;
    float rads, inc, // aux to iterate the polygon
        dist,dist1, // Distance from axis to the rotated center of mass
        Ixx; // What can this be?
    // Principal moments
    float Iu = (b*b*b * apo) / 48;
    float Iv = (b * apo*apo*apo) / 36;
    // Area for the basic triangle
    float area = b * apo / 2;
    // Init iteration
    rads = (inc = 2.0 * M_PI / (float)n) / 2.0;
    Ixx = 0.0;
    // Process all triangles
    for ( i = 0 ; i < n ; i++ ) {
            // Distance from axis to the rotated center of mass
        dist = apo * sin(rads) * 2.0 / 3.0;
        dist1 = 2 * dist;
        // Moment with the rotation and traslation
        float moi = 2*(Iu*cos(rads)*cos(rads) + Iv * sin(rads)*sin(rads));
```

Mechanical Symmetry

```
      float steiner = area * dist*dist + area * dist1*dist1;
      Ixx += moi + steiner;
      rads += inc;
   }
   return( Ixx );
}
float poly_moi_generic () { // Generic formula for CLOSED polygons
   float rads, rads1, inc, y0, x0, x1, y1;
   int i;
   float Ixx; // What can this be?
   // Init iteration
   y0 = 0;
   x0 = R;
   inc = 2 * M_PI / n;
   Ixx = 0.0;
   // Process all vertex    ( order change sign )
   // Note we play with point orientation & order to increase or decrease
   for ( i = 1 ;  i <= n ; i++ ) {
      rads = inc * i;
      rads1 = inc * ( i - 0.5 );
      x1 = 2.0 * apo * cos(rads1);
      y1 = 2.0 * apo * sin(rads1);
      Ixx -= (x1-x0)*(y1+y0)*(y1*y1+y0*y0);
      x0 = R * cos(rads);
      y0 = R * sin(rads);
      Ixx -= (x0-x1)*(y1+y0)*(y1*y1+y0*y0);
   }
   return( Ixx / 12.0 );
}
```

Mechanical Symmetry

Appendix 4: MoI Calculation for any Polygon

Mechanical Symmetry

The formula we are developing is able to calculate the MoI for any polygon, conditioned to be a closed one. Some authors limit its validity to convex polygons. We do not want to (and cannot) overrule that limitation, but this formula has been used to calculate the MoI for stars based on regular polygons.

In the following figure, we see how to break down a polygon (a triangle, to keep it simple) to get the MoI as the sum of the MoI of trapezoids formed by the polygon sides and the moments axis.

Note that the total MoI is the sum; the orientation of the polygon sides segments, gives a sign to each partial MoI. For increasing x values, we will get a positive MoI value, and vice versa.

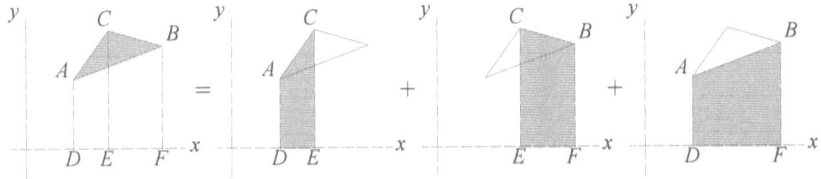

Fig. 41 Polygon as a Sum of Trapezoids

$$I_{ABC} = I_{CADE} + I_{BCEF} + I_{ABFD} \qquad [A4.1]$$

Eq. 35 Polygon as a Sum of Trapezoids – Total MoI

Let's see now how to calculate the MoI for each trapezoid. There are (at least) two ways to get the formula: based on the definition of MoI itself, and more sophisticated, based on Green's theorem.

Ap. 4.1 Trapezoid MoI Sums

$$I_{xi} = \int_\Omega dist^2 \cdot d\Omega = \int_{x_i}^{x_{i+1}} dI$$

$$dI = \frac{y^3}{3} dx$$

Straight line passing both points:

$$y = \frac{y_i - y_{i+1}}{x_i - x_{i+1}} x - \frac{x_{i+1} y_i - x_i y_{i+1}}{x_i - x_{i+1}}$$

Then

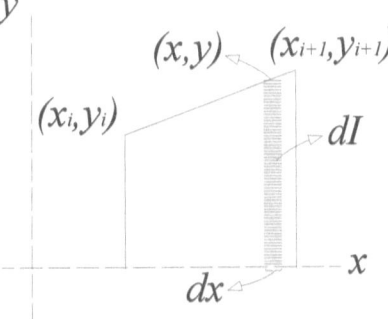

Fig. 42 Trapezoid MoI Respect to Its Basis

$$dI = \frac{1}{3}\left(\frac{y_i - y_{i+1}}{x_i - x_{i+1}}x - \frac{x_{i+1}y_i - x_i y_{i+1}}{x_i - x_{i+1}}\right)^3 \rightarrow$$

[A4.2] $$\rightarrow I_{xi} = \int_{x_i}^{x_{i+1}} dI = \frac{(x_{i+1} - x_i)(y_{i+1} + y_i)(y_{i+1}^2 + y_i^2)}{12}$$

Formula as a function of trapezoid's upper side end points.

For a polygon with n sides:

[A4.3] $$I_x = \sum_{1}^{n} I_{xi} = \frac{1}{12}\sum_{1}^{n}(x_{i+1} - x_i)(y_{i+1} + y_i)(y_{i+1}^2 + y_i^2)$$

Ap. 4.2 Green Theorem

Using this theorem, we can calculate the value for an integral extended to a region D enclosed by a curve C with an integral extended to the line C.

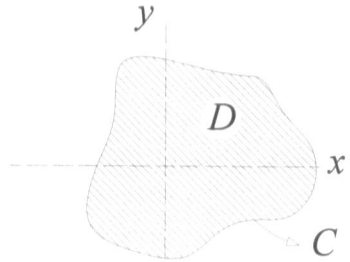

Fig. 43 Green Theorem

$$\oint_C (Ldx + Mdy) = \iint_D \left(\frac{\partial M}{\partial x} - \frac{\partial L}{\partial y}\right) dxdy$$

Eq. 36 Green Theorem

For example, the area:

$$\Omega = \iint_D d\Omega = \iint_D dxdy$$

Then any function fulfilling:

$$\left(\frac{\partial M}{\partial x} - \frac{\partial L}{\partial y}\right) = 1$$

Can be used with the theorem:

$$\left(\frac{\partial M}{\partial x} - \frac{\partial L}{\partial y}\right) = 1 \Rightarrow \begin{cases} M = x & L = 0 \\ M = 0 & L = -y \end{cases} \Rightarrow \oint_C (Ldx + Mdy) = \begin{cases} \oint_C xdy \\ \oint_C -ydx \end{cases}$$

Going back to MoI, specifically to the moment respect x axis:

$$I_x = \iint_D y^2 d\Omega = \iint_D y^2 dxdy$$

Then any function fulfilling

$$\left(\frac{\partial M}{\partial x} - \frac{\partial L}{\partial y}\right) = y^2$$

can be used with the theorem. Let's see:

$$\begin{cases} \left(\dfrac{\partial M}{\partial x} - \dfrac{\partial L}{\partial y}\right) = y^2 \\ M = xy^2 \quad L = 0 \\ M = 0 \quad L = \dfrac{y^3}{3} \end{cases} \Rightarrow \oint_C (Ldx + Mdy) = \begin{cases} \oint_C xy^2 dy \\ \oint_C \dfrac{y^3}{3} dx \end{cases}$$

We see now a clear similarity with the previous system:

$$dI = \frac{y^3}{3} dx \to \oint \frac{y^3}{3} dx$$

Using this on one side of the polygon:

$$I_{xi} = \oint_C \frac{y^3}{3} dx =$$

$$= \left(\frac{-1}{3} \frac{x(x_{i+1}y_i - x_i y_{i+1})^3}{(x_i - x_{i+1})^3} + \frac{3x^2(y_i - y_{i+1})(x_{i+1}y_i - x_i y_{i+1})^2}{2(x_i - x_{i+1})^3}\right.$$

$$\left. -\frac{x^3(y_i - y_{i+1})^2(x_{i+1}y_i - x_i y_{i+1})}{(x_i - x_{i+1})^3} + \frac{x^4(y_i - y_{i+1})^3}{4(x_i - x_{i+1})^3}\right)\Bigg]_{x_i}^{x_{i+1}} =$$

$$= \frac{(x_{i+1} - x_i)(y_{i+1} + y_i)(y_{i+1}^2 + y_i^2)}{12} \qquad [A4.4]$$

With [A4.2] and [A4.4] identical, we have again for the complete polygon:

[A4.3] $$I_x = \sum_1^n I_{xi} = \frac{1}{12}\sum_1^n (x_{i+1} - x_i)(y_{i+1} + y_i)(y_{i+1}^2 + y_i^2)$$

Similarly, for the moment with respect to axis y:

$$I_y = \iint_D x^2 d\Omega = \iint_D x^2 dxdy$$

$$\begin{cases} \left(\dfrac{\partial M}{\partial x} - \dfrac{\partial L}{\partial y}\right) = x^2 \\ M = 0 \quad L = -x^2 y \\ M = \dfrac{x^3}{3} \quad L = 0 \end{cases} \Rightarrow \oint_C (Ldx + Mdy) = \begin{cases} \oint_C -x^2 y dx \\ \oint_C \dfrac{x^3}{3} dy \end{cases}$$

[A4.5] $$I_{yi} = \oint_C \frac{x^3}{3} dy = \frac{(y_{i+1} - y_i)(x_{i+1} + x_i)(x_{i+1}^2 + x_i^2)}{12}$$

For the complete polygon:

[A4.6] $$I_y = \sum_1^n I_{yi} = \frac{1}{12}\sum_1^n (y_{i+1} - y_i)(x_{i+1} + x_i)(x_{i+1}^2 + x_i^2)$$

Similarly, we get I_{xy}:

$$I_{xy} = \iint_D xy d\Omega = \iint_D xy dxdy$$

$$\begin{cases} \left(\dfrac{\partial M}{\partial x} - \dfrac{\partial L}{\partial y}\right) = xy \\ M = 0 \quad L = -x\dfrac{y^2}{2} \\ M = \dfrac{x^2}{2} y \quad L = 0 \end{cases} \Rightarrow \oint_C (Ldx + Mdy) = \begin{cases} \oint_C -x\dfrac{y^2}{2} dx \\ \oint_C \dfrac{x^2}{2} ydy \end{cases}$$

$$I_{xyi} = \oint_C -x\frac{y^2}{2}dx = \quad [A4.7]$$

$$\frac{(x_i - x_{i+1})(3x_{i+1}y_{i+1}^2 + x_iy_{i+1}^2 + 2x_{i+1}y_iy_{i+1} + 2x_iy_iy_{i+1} + x_{i+1}y_i^2 + 3x_iy_i^2)}{24}$$

For the complete polygon:

$$I_{xy} = \sum_1^n I_{xyi} = \quad [A4.8]$$

$$\sum_1^n \frac{(x_i - x_{i+1})(3x_{i+1}y_{i+1}^2 + x_iy_{i+1}^2 + 2x_{i+1}y_iy_{i+1} + 2x_iy_iy_{i+1} + x_{i+1}y_i^2 + 3x_iy_i^2)}{24}$$

Having now our formulas.
But integrating over y:

$$I_{xyi} = \oint_C \frac{x^2}{2} y\, dy = \quad [A4.9]$$

$$\frac{(y_{i+1} - y_i)(3y_{i+1}x_{i+1}^2 + y_ix_{i+1}^2 + 2y_{i+1}x_ix_{i+1} + 2y_ix_ix_{i+1} + y_{i+1}x_i^2 + 3y_ix_i^2)}{24}$$

we could think that [A4.7] and [A4.9] should be equal—but they are not. To be equal, it should fulfilled:

$$[A4.7]-[A4.9] = 0 \implies x_i^2 y_i^2 - x_{i+1}^2 y_{i+1}^2 = 0$$

Similarly for I_x and I_y:

$$I_x \rightarrow x_i y_i^3 - x_{i+1} y_{i+1}^3 = 0$$
$$I_y \rightarrow y_i x_i^3 - y_{i+1} x_{i+1}^3 = 0$$

These three conditions take us to the previously cited condition (included in Green's theorem), asking for a closed line to make the integral along it. For a closed line:

$$x_{n+1} = x_1$$
$$y_{n+1} = y_1$$

and then

$$x_{n+1}y_{n+1} = x_1 y_1$$
$$x_{n+1}y_{n+1}^3 = x_1 y_1^3$$
$$y_{n+1}^2 x_{n+1}^2 = y_1^2 x_1^2$$

Making the total sum for all the polygon sides null:

$$\text{For } I_x \to \sum_1^n x_i y_i^3 - x_{i+1} y_{i+1}^3 = 0$$

$$\text{For } I_y \to \sum_1^n y_i x_i^3 - y_{i+1} x_{i+1}^3 = 0$$

$$\text{For } I_{xy} \to \sum_1^n y_i^2 x_i^2 - y_{i+1}^2 x_{i+1}^2 = 0$$

We confirm that the condition for a closed line is a requirement for the validity of the formulas:

$$[A4.3]\ I_x = \frac{1}{12}\sum_1^n (x_{i+1} - x_i)(y_{i+1} + y_i)(y_{i+1}^2 + y_i^2)$$

$$[A4.6]\ I_y = \frac{1}{12}\sum_1^n (y_{i+1} - y_i)(x_{i+1} + x_i)(x_{i+1}^2 + x_i^2)$$

$$[A4.8]\ I_{xy} = \sum_1^n \frac{(x_i - x_{i+1})(3x_{i+1}y_{i+1}^2 + x_i y_{i+1}^2 + 2x_{i+1} y_i y_{i+1} + 2x_i y_i y_{i+1} + x_{i+1} y_i^2 + 3x_i y_i^2)}{24}$$

Eq. 37 Formulas for MoI calculation for any polygon

Mechanical Symmetry

Mechanical Symmetry

Index of Figures

1 Torque 3
2 Solid Moment of Inertia 4
3 Particles Moment of Inertia 4
4 Steiner's Theorem 4
5 Axial Symmetry 5
6 Point Symmetry 5
7 Rotational Symmetry 5
8 Axial Symmetry - Figures 5
9 Axial Symmetry - Functions 5
10 Point Symmetry - Figure 6
11 Point Symmetry - Functions 6
12 Rotational Symmetry - Figures 7
13 Function with Rotational Symmetry 7
14 Steiner - Section and Axis Translation 19
15 Steiner - Correct calculation - Axis not at center of mass 20
16 MoI Rotation – Section and Axis Rotation 21
17 Section Rotation Formulas 22
18 Superposition and Distance to Axis Influence on MoI 24
19 Particles MoI - Summations 37
20 Particles MoI - Proof 39
21 Thin Wall Tube MoI – MS Formula 45
22 Thin Wall Tube MoI 45
23 MoI Generic Section 46
24 Mechanical Symmetry without Rotational Symmetry 51
25 Necessary Condition for Mechanical Symmetry –
  Even Function 54
26 Necessary Condition for Mechanical Symmetry –
  Rotated Even Function 54
27 Incoherent Sums 66
28 Regular Polygons 71
29 Circles and Computers – Discretization 76
30 Circles and Computers – Conditions 1 & 2 80
31 Circles and Computers – Conditions 1 & 2 (detail) 80
32 Circles and Computers – Conditions 2 & 3 82
33 Circles and Computers – Conditions 2 & 3 (detail) 83
34 Circles and Computers – Conditions 1, 2 & 3 86
35 Program STEINER –
  MoI Change Visualization (B&W Version) 139
36 Sin^2 Sums (Black and White Version from Color Original) 157
37 MoI Graphical Interactive Comparison (B&W version) 175

38 Regular Polygons MoI Interactive Calculation
  (B&W version)... 193
39 Regular Polygonal Tubes MoI Interactive Calculation
  (B&W version)... 207
40 Regular Polygon Based Stars MoI Interactive Calculation
  (B&W version)... 211
41 Polygon as a Sum of Trapezoids 217
42 Trapezoid MoI Respect to Its Basis 217
43 Green Theorem.. 218

Index of Equations

1 Torque 3
2 Solid Moment of Inertia 4
3 Particles Moment of Inertia 4
4 Steiner's Theorem 4
5 Moment of Inertia – Rotation 17
6 Moment of Inertia - Principal Directions 18
7 Constant Moment of Inertia 18
8 Particles MoI – First Summation 37
9 Particles MoI – Variable Summand 37
10 Particles MoI – Approximate Formula 38
11 Particles MoI – Exact Formula 38
12 Euler's notation 40
13 MoI Particles – Null Sum Proof 42
14 Particles MoI – Exact Formula Proof 42
15 MoI Particles - Exact 43
16 MoI Particles - Approximate 43
17 MoI Particles Accuracy 43
18 Thin Wall Tube MoI – MS Formula 45
19 Thin Wall Tube MoI 45
20 MoI Generic Section with Mechanical Symmetry. 47
21 MoI Generic Section with MS – Approximate 47
22 MoI Generic Section with MS – Accuracy 47
23 Necessary Condition for Mechanical Symmetry 54
24 Additional Necessary Conditions for Mechanical Symmetry 54
25 Fourier's Series 55
26 Necessary Condition for Mechanical Symmetry – Fourier 56
27 Regular Polygon MoI - From R and k 74
28 Regular Polygon MoI - From ap=h and k 74
29 Regular Polygon MoI - From L and k 74
30 Regular Polygon MoI – With Area 75
31 Circles and Computers – Condition 1 Equal Area 76
32 Circles and Computers – Condition 2 Equal MoI 76
33 Circles and Computers – Condition 3 Equal W Modulus 77
34 Regular Polygons Resolution 96
35 Polygon as a Sum of Trapezoids – Total MoI 217
36 Green Theorem 218
37 Formulas for MoI calculation for any polygon 222

Mechanical Symmetry

Index of Tables

1 Particles MoI – Approximate formula accuracy 44
2 Known Values for the Circle 48
3 Known Values for Quarter of a Circle 48
4 Fourier – Harmonic functions factors 57
5 Fourier – Rotational Symmetry Samples 61
6 Circles and Computers – Individual Factors 87
7 Circles and Computers – Combined Factors and Errors 88
8 Output – Particles MoI Comparison 172

Glossary

Area
: 3, 4, 12, 24, 25, 29, 36, 39, 48, 51, 52, 72, 73, 74, 75, 77, 78, 83, 87, 93, 139, 142, 143, 165, 168, 172, 173, 181, 193, 199, 202, 207, 208, 209, 211, 212, 218

Axial
: 5, 6, 76, 78, 86

Axis
: 3, 4, 5, 12, 19, 20, 22, 23, 24, 26, 27, 36, 37, 39, 51, 54, 61, 76, 95, 165, 167, 199, 208, 209, 212, 217, 219, 220

Coordinates
: 7, 19, 22, 52, 142, 146, 176, 18, 200, 201, 209

Distance
: 3, 4, 23, 24, 35, 36, 72, 73, 74, 76, 165, 167, 199, 208, 209, 212

Exercise
: 8, 25, 47

Formula
: 3, 4, 17, 18, 27, 45, 48, 71, 73, 74, 76, 95, 96, 165, 168, 169, 173, 213, 217, 218, 221, 222

Functions
: 5, 6, 7, 21, 37, 52, 55, 56, 57, 58, 73, 75, 126, 193, 211, 218, 219

Green
: 52, 175, 217, 218, 221

Inertia
: 3, 4, 8, 17, 18, 38, 51, 58, 93, 173

I_x
: 8, 9, 17, 18, 27, 47, 95, 221

I_{xy}
: 17, 18, 20, 26, 27, 48, 52, 95, 220

I_y
: 9, 10, 17, 18, 27, 95, 221

MoI (Moment of Inertia)
: 3, 4, 5, 8, 9, 10, 11, 13, 17, 18, 19, 20, 21, 22, 23, 24, 25, 29, 30, 31, 35, 36, 37, 38, 39, 42, 43, 44, 45, 47, 51, 52, 53, 57, 71, 73, 74, 75, 76, 77, 79, 80, 83, 87, 139, 140, 146, 147, 165, 166, 167, 168, 169, 172, 173, 175, 176, 181, 193, 194, 199, 202, 207, 208, 211, 212, 215, 217, 219, 222

Moment
: 3, 4, 8, 38, 46, 48, 51, 58, 66, 143, 173, 199, 217, 219, 220

Order
: 3, 6, 7, 17, 37, 38, 39, 42, 43, 46, 51, 57, 58, 200, 209, 213

Particle	Section
4, 37, 38, 39, 43, 153, 155, 161, 167, 169, 170, 171, 173, 175, 177, 178, 182	4, 17, 18, 19, 22, 23, 24, 36, 37, 38, 39, 43, 45, 46, 47, 48, 51, 52, 53, 54, 57, 58, 61, 75, 76, 91, 139, 140, 141, 143, 153, 155, 161, 165, 169
Point	
3, 4, 5, 6, 7, 19, 22, 54, 217, 218	
Polygon	Steiner
71, 74, 75, 76, 77, 95, 96, 97, 98, 110, 11, 112, 193, 207, 211, 215, 217, 218, 219, 220, 221, 222	4, 11, 17, 18, 19, 20, 25, 28, 29, 39, 49, 139, 140, 143, 165, 167, 169
Polygonal	Symmetry
113, 115, 116, 117, 122, 123, 124, 125, 126, 127, 129, 131, 132, 133	1, 5, 6, 7, 17, 33, 35, 37, 38, 39, 42, 43, 45, 46, 47, 49, 51, 54, 56, 57, 58, 61, 71, 74, 91, 93, 95, 140, 144, 146, 153, 154, 155, 156, 157, 158, 159, 165, 166, 167, 169, 176, 180, 181, 194, 199, 201, 202, 208, 212
Principal (direction or moment)	
18, 26, 46, 47, 48, 61, 93	
Program	
38, 137, 139, 140, 151, 153, 154, 157, 165, 166, 169, 175, 176, 186, 191, 194, 201, 207, 208, 211, 212	Test
	35
	Theorem
	4, 17, 18, 19, 20, 23, 29, 49, 51, 52, 58, 140, 217, 218, 219
Property(ies)	
3, 5, 17, 76, 86, 87	Torque
Rotation	3
6, 17, 21, 22, 25, 27, 37, 38, 54, 55, 58, 93, 139, 140, 142, 143, 155, 161, 199, 208, 209, 212	Translation
	5, 6, 19, 20, 25, 28, 139
Rotational	
5, 7, 17, 37, 39, 42, 43, 45, 46, 51, 57, 58, 61, 71	

Bibliography

[1] Pisarenko, G. S., Yákovlev, A. P., Matviéev, V. V. Manual de Resistencia de Materiales. URSS. 1975. ISBN 5-88417-035-1.

[2] Ruiz-Tolosa, J. R. *Caminando de los Vectores a los Tensores.* Academia de Ingeniería. 2006. ISBN 84-95662-31-0.

[3] Kernighan, B. W., Ritchie, D. M. *El Lenguaje de Programación C.* Prenctice-Hall Iberoamericana. 1985. ISBN 968-880-024-4.

Mechanical Symmetry

Acknowledgements

To write and prepare this book, we have used tools from:

- Maxima, Computer Algebra System: http://maxima.sourceforge.net/
- WxMaxima, Document Based Interface: http://andrejv.github.com/wxmaxima/
- Chikrii Softlab: http://www.chikrii.com/products/tex2word
- BloddShed Software Dev-CPP: http://www.bloodshed.net/devcpp.html
- Decimal Basic: http://hp.vector.co.jp/authors/VA008683/english/

For their contributions to my work:

To my family.

To my teachers.

To my friends and colleagues.

www.ingramcontent.com/pod-product-compliance
Lightning Source LLC
Chambersburg PA
CBHW020745180526
45163CB00001B/360